Applied Statistics

A HANDBOOK OF BMDP™ ANALYSES

Applied Statistics
A HANDBOOK OF BMDP™ ANALYSES

E. J. SNELL
Department of Mathematics
Imperial College
London

A complement to

Applied Statistics
PRINCIPLES AND EXAMPLES

D. R. COX AND E. J. SNELL
Department of Mathematics
Imperial College
London

LONDON   NEW YORK
CHAPMAN AND HALL

First published in 1987 by
Chapman and Hall Ltd
11 New Fetter Lane, London EC4P 4EE

Published in the USA by
Chapman and Hall
29 West 35th Street, New York NY 10001

© 1987 E.J. Snell

Printed in Great Britain by
J.W. Arrowsmith Ltd, Bristol

ISBN 0 412 28410 3

British Library Cataloguing in Publication Data

Snell, F.J.
    Applied statistics: a handbook of BMDP
    analyses
    1. BMDP (Computer program)
    I. Title II. Cox, D.R. Applied statistics
    519.5'028'55329   QA276.4

    ISBN 0-412-28410-3

Library of Congress Cataloging in Publication Data

Snell, E. J.
    Applied statistics.

    "A complement to Applied statistics: principles and
examples [by] D.R. Cox and E.J. Snell."
    Bibliography: p.
    Includes index.
    1. Mathematical statistics. 2. Medical statistics—
Data processing. I. Cox, D. R. (David Roxbee). Applied
statistics. II. Title.
    QA276.C6653 1981 Suppl.   519.5   86–21529
    ISBN 0–412–28410–3

Contents

Foreword

This handbook is a realization of a long term goal of BMDP Statistical Software. As the software supporting statistical analysis has grown in breadth and depth to the point where it can serve many of the needs of accomplished statisticians it can also serve as an essential support to those needing to expand their knowledge of statistical applications. Statisticians should not be handicapped by heavy computation or by the lack of needed options.

When *Applied Statistics, Principle and Examples* by Cox and Snell appeared we at BMDP were impressed with the scope of the applications discussed and felt that many statisticians eager to expand their capabilities in handling such problems could profit from having the solutions carried further, to get them started and guided to a more advanced level in problem solving. Who would be better to undertake that task than the authors of *Applied Statistics*?

A year or two later discussions with David Cox and Joyce Snell at Imperial College indicated that a wedding of the problem statements and suggested solutions with control language to accomplish these analyses would further the learning process for many statisticians. They were willing to undertake the project.

Joyce Snell has done an excellent job of melding the two approaches and has carried many of the problems a step further by suggesting alternate approaches and follow-up analyses.

W. J. Dixon
BMDP Statistical Software

Preface

Examples covering a wide range of statistical techniques are discussed in *Applied Statistics. Principles and Examples* (Cox and Snell, 1981). Attention there concentrates, not so much on computational details, but more on general methodological issues and the interpretation of conclusions. It is therefore recommended that students should work through the detailed analyses themselves, to obtain the results discussed and examine the conclusions from alternative or more extensive analyses. Even though some of the sets of data are small enough for hand analysis, a computer can be helpful, particularly when a complex or repetitive analysis is required.

The BMDP computer programs provide methods ranging from simple plots and descriptive measures to the fitting of complex models. The present handbook lists instructions for using BMDP programs to analyse all the examples discussed in the previous book. The programs are shown to be useful for both non-standard and standard statistical problems.

I am greatly indebted to Dr. W. J. Dixon for suggesting the preparation of this handbook and for providing a Statcat microcomputer with BMDP statistical software which made the task possible. I am most grateful also to Linda Muthen, MaryAnn Hill and Jennifer Row of BMDP for their invaluable help and guidance throughout the project and for preparing camera-ready laser copies of the input and output. Finally I express my appreciation to Sir David Cox for many helpful suggestions and for my earlier pleasure in being a joint author with him of *Applied Statistics. Principles and Examples*, which provides the basis for this handbook.

E. J. Snell
London, May 1986

INTRODUCTION

Introduction

Instructions are given to enable the student to use BMDP statistical computer programs to analyse the examples discussed in *Applied Statistics. Principles and Examples* (Cox and Snell, 1981), hereafter referred to simply as *App. Stat.*

The aim throughout this handbook has been, as far as possible, to keep the BMDP instructions simple yet to use the flexibility of the BMDP programs to provide useful output and in particular to be able to reproduce the tables and results discussed in *App. Stat.* The examples are labelled Example A, . . . , X as in that book and tables of results, etc., given there are referred to, for example, as Table A.2, *App. Stat.*

Each set of data is reproduced in full, with a description of the data, exactly as in *App. Stat.* in order to make this handbook, to some extent, self-contained although it must be used in parallel with *App. Stat.* if full benefit from an analysis is to be gained. Not all tables of results are reproduced here. Selected items from the BMDP output (labelled Output A.1, . . .) are shown and are sufficient for the student to see that the relevant tables or results can be obtained.

BMDP instructions are given for each Example A, . . . , X and also for two further data sets (Sets 3, 14; *App. Stat.*), the latter being selected as illustrating the application of two BMDP programs for techniques not used in the main examples. Notes are given beside the instructions as a help to students.

The statistical analyses range from very simple, barely justifying the use of a computer, to complex, but for the sake of completeness each Example A, . . . , X has been included. Example E is particularly simple, requiring only plots but, as such, provides an easy starting point for anyone wishing for an introduction to the use of BMDP programs. Likewise Example D, although containing interesting discussion in *App. Stat.*, here computes only a simple regression and plots residuals. Example P involves straightforward multiple regression. Other more complex problems involve the fitting of logistic models (Examples H, X), log-linear models (Example W), survival analysis (Set 3), cluster analysis (Set 14) and specific models fitted by maximum likelihood (Examples T, U). Example C provides a nonstandard analysis.

In a course of instruction, the order in which the examples should be taken will depend partly on the need to run simple programs first and partly on the order in which statistical topics are discussed. Readers interested in using BMDP for particular statistical applications, e.g. multiple regression, may go straight to the relevant examples.

Sometimes more than one BMDP program is suitable for a particular example, with one perhaps preferred for some aspects of the analysis. In such cases either instructions are given for the alternative program or it is simply presumed that a similar application in a different example will provide sufficient guidance. A reference list of the BMDP programs used (or suggested) for each example is given in the following section.

The BMDP instructions listed here have not been written by an expert in statistical computing or by someone with long experience of BMDP. The aim has been to keep things at a simple level and to obtain the numerical results quoted in *App. Stat.* in a way that a new user will find straightforward, natural and appealing. Sometimes other analyses are also suggested. Reference to the BMDP manual (Dixon *et al.*, 1985) and user's digest (Hill, 1984) will undoubtedly prove valuable. Some general points on the BMDP instructions as used in the examples are given in the subsequent section.

The programs given in this handbook have been run using the 1985 release of BMDP Statistical Software.

BMDP programs used (or suggested) for the examples

Example	BMDP programs	Example	BMDP programs
A	2D	N	9D, 4F, 2V, (4V, 8V)
B	1D, 6D	O	2V
C	1D	P	1R, 9R
D	1R, (2R)	Q	7D, 8V, (3V)
E	6D	R	8V, (2V)
F	1R, (2R)	S	2V
G	1R, 2R, 9R	T	AR, (3R)
H	3R, LR, (AR)	U	3R, (AR)
I	1R, (2R)	V	LR
J	9D, 1R, (2R)	W	4F
K	4V, (9D, 2V, 8V, 1R, 2R)	X	4F, LR
L	4V, LR, (2V, 1R, 2R)	Set 3	1L, 2L
M	1R, 2R, LR	Set 14	1M, (4M)

(Instructions are not given for programs shown in brackets.)

Program descriptions
1D Simple data description
2D Detailed data description, including frequencies
6D Bivariate (scatter) plots
7D Description of groups (strata) with histograms and analysis of variance
9D Multiway description of groups
4F Two-way and multiway frequency tables – measures of association and the log-linear model (complete and incomplete tables)
1L Life tables and survival functions
2L Survival analysis with covariates – Cox models
1M Cluster analysis of variables

4M Factor analysis
1R   Multiple linear regression
2R   Stepwise regression
3R   Nonlinear regression
9R   All possible subsets regression
AR   Derivative-free nonlinear regression
LR   Stepwise logistic regression
2V   Analysis of variance and covariance, including repeated measures
3V   General mixed model analysis of variance
4V   General univariate and multivariate analysis of variance and covariance, including repeated measures
8V   General mixed model analysis of variance – equal cell sizes

Some points on the BMDP instructions

For a detailed explanation on any point, reference should be made to the BMDP manual (Dixon *et al.*, 1985) or to the BMDP user's digest (Hill, 1984). These notes are intended only as a brief guide on the BMDP instructions as used in this handbook.

1. System instructions need to be prefixed to the BMDP instructions given for each example. The format for system instructions will depend on the particular installation or system being used. BMDP programs are available for systems ranging from mainframe to micro or personal computer.

2. All the sets of data given in the examples are fairly small. Even so, on small systems it may be necessary to increase the program capacity (which can usually be done) or to run the analysis in segments (see Example N).

3. Some users may have access to only a selection of BMDP programs, in which case alternative programs to those illustrated can probably be used. Several of the examples can be analysed by multiple regression (BMDP1R or BMDP2R) or by maximum likelihood (BMDP3R, BMDPAR) in place of the methods described.

4. Standard BMDP language, e.g. 'paragraphs' and 'sentences', is used in describing instructions.

5. Instructions for all examples start with the three paragraphs:

   /problem
   /input
   /variable

   The **problem** paragraph is optional if the **input** paragraph is used but it is useful for labelling the output. Thus

   /problem   title = 'example b. birth intervals. bmdp1d'.

   causes the output to be labelled with the title within quotes.

7

The **input** paragraph is obligatory for the first problem, specifying how many variables are to be read and the form of the input. We have chosen always to use free format (the data being separated by blanks or commas and each record starting on a new line) although stream, slash or fixed format are permissible.

The **variable** paragraph is recommended although is not always essential. The **names** list in this paragraph identifies the order for input of the data and provides labels for the output. Most of the examples require variables to be generated in a **transform** paragraph and the names of these variables can be automatically added to those already specified by stating **add = new.** in the **variable** paragraph. For example, in Example I we have

/variable   names = c, h, a, t, v, l, y1, y2.
              add = new.
/transform   w = y2 − y1.

where **w** is generated in order to be used in the analysis. **add = new.** accommodates the additional variable. **names = c, . . . , y2** specifies the order in which the data for each record are to be read.

Names not beginning with a letter, or names which include nonnumeric or nonalphabetic symbols, need to be enclosed in single quotes. Such names have been avoided in the examples (except in Example N). Names containing more than 8 characters are automatically shortened by the program to use the first 8 characters.

6. Some BMDP programs repeat an analysis for each variable of the **variable** paragraph (including any added by **add = new.**). The **use** statement will restrict the choice to a subset of the variables, e.g. in Example A we have

use = dwk, time, date, hours.

which causes the program to form frequency tables for only these four variables.

7. For reasons of space, only the first two lines of data are shown following the instructions for each example. The data set should be concluded with **/end**, e.g. the full data for Example F read as follows:

```
1 1   5.84
2 2   5.73
3 2   7.30
4 1  10.46
5 1   9.71
6 2   5.91
/end
```

although in practice programs will run without **/end** (but will print a pseudo-error message).

8. Many of the examples call for more than one approach to the analysis, using different programs, and hence require more than one submission of the data. To ease the preparation of input we have deliberately therefore kept the same data format throughout an example (except in Examples Q and R where BMDP8V requires a special format). It can often be helpful to have the data on a separate file, as in 9(ii) below.

9. Some examples require the same form of analysis to be repeated for subsets of the data, e.g. Example S requires BMDP2V to be used for each of the 10 setting stations. This can be achieved in two ways:
(i) submit the program and full data set 10 times and each time include a sentence in the **transform** paragraph specifying which setting station to use, e.g.

    /transform use = sstation eq 1.

will analyse the data for only setting station 1; or
(ii) use the **for–%** option, but note that when this option is used to repeat a whole problem (rather than repeat instructions within a paragraph, as in Example I) it is necessary to have the data on a separate file and to include the name of the file in the **input** paragraph, e.g. after the **problem** paragraph write

    for k = 1 to 10.
% /input variables = 4.
        format = free.
        file = 'datas'.

and follow with the **variable** paragraph; file **datas** contains the 256 lines of data for Example S. Operational details may differ slightly according to installation; see BMDP manual (Dixon, 1985). Also amend the sentence in the **transform** paragraph to **use = sstation eq k.**

10. Points which are more specific to particular programs are mentioned in the examples as they arise.

EXAMPLES

Example A

Admissions to intensive care unit

Description of data

Table A.1 gives arrival times of patients at an intensive care unit. The data were collected by Dr A. Barr, Oxford Regional Hospital Board. Interest lies in any systematic variations in arrival rate, especially any that might be relevant in planning future administration.

Table A.1. Arrival times of patients at intensive care units (To be read down the columns.)

1963			1963			1963			1963		
M	4 Feb.	11.00 hr			22.00hr			12.20hr			23.55hr
		17.00	S	6 Apr.	22.05	M	3 June	14.45	S	20 July	3.15
F	8 Feb.	23.15	T	9 Apr.	12.45	W	5 June	22.30	Su	21 July	19.00
M	11 Feb.	10.00			19.30	M	10 June	12.30	T	23 July	21.45
S	16 Feb.	12.00	W	10 Apr.	18.45			13.15	W	24 July	21.30
M	18 Feb.	8.45	Th	11 Apr.	16.15	W	12 June	17.30	S	27 July	0.45
		16.0	M	15 Apr.	16.00	Th	13 June	11.20			2.30
W	20 Feb.	10.00	T	16 Apr.	20.30			17.30	M	29 July	15.30
		15.30	T	23 Apr.	23.40	Su	16 June	23.00	Th	1 Aug.	21.00
Th	21 Feb.	20.20	Su	28 Apr.	20.20	T	18 June	10.55	F	2 Aug.	8.45
M	25 Feb.	4.00	M	29 Apr.	18.45			13.30	S	3 Aug.	14.30
		12.00	S	4 May	16.30	F	21 June	11.00			17.00
Th	28 Feb.	2.20	M	6 May	22.00			18.30	W	7 Aug.	3.30
F	1 Mar.	12.00	T	7 May	8.45	S	22 June	11.05			15.45
Su	3 Mar.	5.30	S	11 May	19.15	M	24 June	4.00			17.30
Th	7 Mar	7.30	M	13 May	15.30			7.30	Su	11 Aug.	14.00
		12.00	T	14 May	12.00	Tu	25 June	20.00	T	13 Aug.	2.00
S	9 Mar.	16.00			18.15			21.30			11.30
F	15 Mar.	16.00	Th	16 May	14.00	W	26 June	6.30			17.30
S	16 Mar.	1.30	S	18 May	13.00	Th	27 June	17.30	M	19 Aug.	17.10
Su	17 Mar.	11.05	Su	19 May	23.00	S	29 June	20.45	W	21 Aug.	21.20
W	20 Mar.	16.00	M	20 May	19.15	Su	30 June	22.00	S	24 Aug.	3.00
F	22 Mar.	19.00	W	22 May	22.00	T	2 July	20.15	S	31 Aug.	13.30
Su	24 Mar.	17.45	Th	23 May	10.15			21.00	M	2 Sept.	23.00
		20.20			12.30	M	8 July	17.30	Th	5 Sept.	20.10
		21.00	F	24 May	18.15	T	9 July	19.50	S	7 Sept.	23.15
Th	28 Mar.	12.00	S	25 May	21.05	W	10 July	2.00	Su	8 Sept.	20.00
		12.00	T	28 May	21.00	F	12 July	1.45	T	10 Sept.	16.00
S	30 Mar.	18.00	Th	30 May	0.30	S	13 July	3.40			18.30
T	2 Apr.	22.00	S	1 June	1.45			4.15			*continued*

13

1963	1963	1963–64	1964
W 11 Sept. 21.00hr	Th 7 Nov. 13.30hr	W 11 Dec. 14.00hr	F 17 Jan. 16.40hr
F 13 Sept. 21.10	F 8 Nov. 12.30	Th 12 Dec. 21.15	Su 19 Jan. 18.00
Su 15 Sept. 17.00	S 9 Nov. 13.45	F 13 Dec. 18.45	M 20 Jan. 20.00
M 16 Sept. 13.25	19.30	S 14 Dec. 14.05	T 21 Jan. 11.15
W 18 Sept. 15.05	M 11 Nov. 0.15	14.15	F 24 Jan. 16.40
S 21 Sept. 14.10	T 12 Nov. 7.45	Su 15 Dec. 1.15	S 25 Jan. 13.55
M 23 Sept. 19.15	F 15 Nov. 15.20	M 16 Dec. 1.45	W 29 Jan. 21.00
T 24 Sept. 14.05	18.40	T 17 Dec. 18.00	Th 30 Jan. 7.45
22.40	19.50	F 20 Dec. 14.15	F 31 Jan. 22.30
F 27 Sept. 9.30	S 16 Nov. 23.55	15.15	W 5 Feb. 16.40
S 28 Sept. 17.30	Su 17 Nov. 1.45	S 21 Dec. 16.15	23.10
T 1 Oct. 12.30	M 18 Nov. 10.50	Su 22 Dec. 10.20	Th 6 Feb. 19.15
W 2 Oct. 17.30	T 19 Nov. 7.50	M 23 Dec. 13.35	F 7 Feb. 11.00
Th 3 Oct. 14.30	F 22 Nov. 15.30	17.15	T 11 Feb. 0.15
16.00	S 23 Nov. 18.00	T 24 Dec. 19.50	14.40
Su 6 Oct. 14.10	23.05	22.45	W 12 Feb. 15.45
T 8 Oct. 14.00	Su 24 Nov. 19.30	W 25 Dec. 7.25	M 17 Feb. 12.45
S 12 Oct. 15.30	T 26 Nov. 19.00	17.00	T 18 Feb. 17.00
Su 13 Oct. 4.30	W 27 Nov. 16.10	S 28 Dec. 12.30	18.00
S 19 Oct. 11.50	F 29 Nov. 10.00	T 31 Dec. 23.15	21.45
Su 20 Oct. 11.55	S 30 Nov. 2.30	Th 2 Jan. 10.30	W 19 Feb. 16.00
15.20	22.00	F 3 Jan. 13.45	Th 20 Feb. 12.00
15.40	Su 1 Dec. 21.50	Su 5 Jan. 2.30	Su 23 Feb. 2.30
T 22 Oct. 11.15	M 2 Dec. 19.10	M 6 Jan. 12.00	M 24 Feb. 12.55
W 23 Oct. 2.15	Tu 3 Dec. 11.45	T 7 Jan. 15.45	T 25 Feb. 20.20
S 26 Oct. 11.15	15.45	17.00	W 26 Feb. 10.30
W 30 Oct. 21.30	16.30	17.00	M 2 Mar. 15.50
Th 31 Oct. 3.00	18.30	F 10 Jan. 1.30	W 4 Mar. 17.30
F 1 Nov. 0.40	Th 5 Dec. 10.05	20.15	F 6 Mar. 20.00
10.00	20.00	S 11 Jan. 12.30	T 10 Mar. 2.00
M 4 Nov. 9.45	S 7 Dec. 13.35	Su 12 Jan. 15.40	W 11 Mar. 1.45
23.45	16.45	T 14 Jan. 3.30	W 18 Mar. 1.45
T 5 Nov. 10.00	Su 8 Dec. 2.15	18.35	2.05
W 6 Nov. 7.50	M 9 Dec. 20.30	W 15 Jan. 13.30	

The analysis

It would be simple to calculate the frequency distributions for time-of-day, day-of-week, and calendar-month (Tables A.2, A.3 and A.4, *App. Stat.*) by hand, but calculation of the time intervals between admissions, particularly if Table A.5, *App. Stat.* were extended, is less tedious done by computer. If reading in the data to construct one frequency table, it is sensible to construct others at the same time.

The time interval between successive admissions is very conveniently calculated using the **days** and **retain** features.

Program

BMDP2D is used for calculating all the required frequency distributions.

For day-of-week (Table A.3, *App. Stat.*) the instructions are:

```
/problem     title = 'example a.  hospital
             admissions.  day-of-week.  bmdp2d'.
/input       variables = 5.  format = free.
/variable    names = dwk, day, month, time, year.
             use = dwk.
/end
1 04 02 11.00 63
1 04 02 17.00 63
    . . .
```

use only **dwk.**

254 lines of data
Code: Mon. = 1, Tues. = 2, . . .

Note that the data are entered in the order named in the **variable** paragraph with day preceding month (i.e. British date system). Output A.1 shows the frequency distribution by day-of-week. A histogram (not shown here) is also printed. Various statistics (skewness, mean, standard deviation, etc.) are also printed but are irrelevant in this case and are not shown here.

For time-of-day (Table A2, *App. Stat.*) it is desirable to group the time into intervals, otherwise a large number of different values will be printed. To group into two-hour intervals we can insert a **count** paragraph:

```
/count       truncate = (time)2.
```

This truncates the value of the variable **time** to the nearest lower number divisible by 2. **use = dwk.** in the previous instructions should be changed to **use = time.** The frequency distribution by time-of-day is shown in Output A.2.

Note that using the **truncate** instruction, any statistics such as mean and standard deviation in the printout will have been calculated using the lower class boundaries in place of actual values.

For calendar-month (Table A.4, *App. Stat.*) it is essential to distinguish between, for example, March 1963 and March 1964. We do this by compounding month and year into one variable (e.g. March 1963 = 6303) by inserting a **transform** paragraph:

```
/transform    month = month + year*100.
```

Also **use = month.** must be written in the **variable** paragraph. Output A.3 shows the frequency distribution. The fitted frequencies of Table A.4, *App. Stat.* can be calculated by hand assuming a constant arrival rate of 254 arrivals over 409 days.

Intervals between successive admissions can be calculated using **days** and **retain. days (month, day, year)** expresses the date as the number of days since 1 January 1960 and **retain** causes this number to be held until the next line of data is read. The time of day is converted from hr. (and min.) to hr. (and decimals). It is then straightforward to obtain the interval between admissions (denoted **hours** in the instructions which follow). The intervals are grouped into classes 0 hr–, 2 hr–, 4 hr–, 6 hr or more, corresponding to those of Table A.5, *App. Stat.*; a wider spread of classes would be more informative.

Rather than a separate computer run for each frequency distribution we can compound the instructions into one set, producing all four frequency distributions of Tables A.2–A.5, *App. Stat.* with one submission of the data. The instructions are:

```
/problem     title = 'example a.  hospital
             admissions.  bmdp2d'.
/input       variables = 5.  format = free.
/variable    names = dwk, day, month, time, year.
             add = new.
             use= dwk, time, date, hours.  retain.
/transform   date = month + year*100.              compound month/year
             lagt = timdec.                         lagt retained
             timdec = (time - int(time))/0.60 +     time in hr. (and decimal)
             int(time).
             lnumdays = numdays.                     lnumdays retained
             numdays = days(month,day,year).         days since 1 Jan. 1960
             hours=((numdays - lnumdays)*24) +       interval between submissions
             (timdec - lagt).
             if (hours gt 6) then hours = 6.    ◄—   final class boundary, 6 hr.
/count       truncate = (time)2, (hours)2.     ◄—   2-hr. grouping for time and hours
/end
1 04 02 11.00 63
1 04 02 17.00 63                                     254 lines of data as before
    . . .
```

Output A.4 shows the distribution of time intervals. (Note early printings of *App. Stat.* contain an error in Table A.5).

If the statistics (mean, etc.) which are also printed, but not shown here, are required to be computed using individual, rather than grouped values, an additional run must be made removing the two instructions marked with arrows; in this case printing of the tabulations can be suppressed by inserting a **print** paragraph before **/end** as follows:

/print   **no count.**

Suggested further work

(i) For the time intervals between successive admissions:
   (a) obtain the mean and standard deviation using the ungrouped data;
   (b) extend the frequency destination of Output A.4 to make the final class 12 hr. or more.
(ii) When would unequal interval widths be appropriate and how would you compute the frequency distribution with unequal intervals?

Output A.1

Frequency distribution by day-of-week. BMDP2D

		PERCENTS				PERCENTS	
PERCENTS				PERCENTS			
VALUE	COUNT	CELL	CUM	VALUE	COUNT	CELL	CUM
1.	37	14.6	14.6	5.	30	11.8	71.7
2.	53	20.9	35.4	6.	44	17.3	89.0
3.	35	13.8	49.2	7.	28	11.0	100.0
4.	27	10.6	59.8				

Output A2

Frequency distribution by time-of-day. BMDP2D

		PERCENTS				PERCENTS	
PERCENTS				PERCENTS			
VALUE	COUNT	CELL	CUM	VALUE	COUNT	CELL	CUM
0.	14	5.5	5.5	12.	31	12.2	41.3
2.	17	6.7	12.2	14.	30	11.8	53.1
4.	5	2.0	14.2	16.	36	14.2	67.3
6.	8	3.1	17.3	18.	29	11.4	78.7
8.	5	2.0	19.3	20.	31	12.2	90.9
10.	25	9.8	29.1	22.	23	9.1	100.0

Output A.3

Frequency distribution by calendar-month. BMDP2D

		PERCENTS				PERCENTS	
PERCENTS				PERCENTS			
VALUE	COUNT	CELL	CUM	VALUE	COUNT	CELL	CUM
6302.	13	5.1	5.1	6309.	17	6.7	51.2
6303.	16	6.3	11.4	6310.	17	6.7	57.9
6304.	12	4.7	16.1	6311.	28	11.0	68.9
6305.	18	7.1	23.2	6312.	32	12.6	81.5
6306.	23	9.1	32.3	6401.	23	9.1	90.6
6307.	16	6.3	38.6	6402.	17	6.7	97.2
6308.	15	5.9	44.5	6403.	7	2.8	100.0

Output A.4

Frequency distribution of successive time intervals. BMDP2D

		PERCENTS				PERCENTS	
PERCENTS				PERCENTS			
VALUE	COUNT	CELL	CUM	VALUE	COUNT	CELL	CUM
0.	20	7.9	7.9	4.	7	2.8	15.8
2.	13	5.1	13.0	6.	213	84.2	100.0

Example B

Intervals between adjacent births

Description of data

The data in Table B.1 were obtained by Greenberg and White (1963) from the records of a Genealogical Society in Salt Lake City. The distribution of intervals between successive births in a particular serial position is approximately log normal and hence geometric means are given. For example, the entry 39.9 in the bottom right-hand corner means that for families of 6 children, in which the fifth and sixth children are both girls, the geometric mean interval between the births of these two children is 39.9 months.

Table B.1. Mean intervals in months between adjacent births by family size and sequence of sex at specified adjacent births

Family size	Births	Sequence of sex			
		MM	MF	FM	FF
2	1–2	39.8	39.5	39.4	39.3
3	1–2	31.0	31.5	31.4	31.1
3	2–3	42.8	43.7	43.3	43.4
4	1–2	28.4	28.1	27.5	27.8
4	2–3	34.2	34.4	34.3	35.0
4	3–4	43.1	44.3	43.3	42.8
5	1–2	25.3	25.6	25.6	25.5
5	2–3	30.3	30.1	29.9	30.0
5	3–4	33.7	34.0	33.7	34.7
5	4–5	41.6	42.1	41.9	41.3
6	1–2	24.2	24.4	24.0	24.5
6	2–3	27.6	27.7	27.5	27.6
6	3–4	29.8	30.2	30.3	30.8
6	4–5	34.2	34.2	34.1	33.4
6	5–6	40.3	41.0	40.6	39.9

The analysis

The analysis here is at a very descriptive level and the amount of data small so it is doubtful whether a computer run is really justified. However, since the data are geometric means, it is appropriate to take logarithms for the calculations and this can conveniently be done in a **transform** paragraph. Row means can also be plotted for the different family sizes as in Fig. B.2, *App. Stat.*

Programs

BMDP1D is used to obtain for each row of Table B.1 the average birth interval in log months and its antilog (Table B.2, *App. Stat.*), plus the range of log months. The **max**, **min** and **mean** facilities are particularly useful. The instructions are:

```
/problem    title = 'example b.  birth intervals.
            bmdp1d'.
/input      variables = 6.  format = free.
/variable   names = size, birth, mm, mf, fm, ff.
            add = new.
            use = size, range, lmonths, months.
/transform  range = ln(max(mm,mf,fm,ff)) -      range of log months
            ln(min(mm,mf,fm,ff)).
            lmonths = mean(ln(mm),ln(mf),ln(fm),   average log months
            ln(ff)).
            months = exp(lmonths).               antilog of average
/print      data.
/end
2 2 39.8 39.5 39.4 39.3                          15 lines of data
3 2 31.0 31.5 31.4 31.1                          Code: births 1–2 = 2,
                                                       2–3 = 3, etc.
  . . .
```

The output lists values of all the varieties named in the **variable** paragraph for each line of data. Output B.1 shows the average values of log months and corresponding antilogs, as in Table B.2, *App. Stat.*, together with the values of the range of log months. The average range over all 15 lines of data, shown in Output B.2, was used in *App. Stat.* to provide an estimated standard deviation for log months, given that the sequence MM, . . . , FF is ignored.

BMDP6D is used to plot the data, plotting the average of log months against order of birth, aligned according to final interval (0, final interval; −1, previous interval, . . .) as in Fig. B.2, *App. Stat.* The instructions are:

```
/problem    title = 'example b.  birth intervals.
            bmdp6d'.
/input      variables = 6.  format = free.
/variables  names = size, birth, mm, mf, fm, ff.
            add = new.  group = size.           identify size on plot.
/transform  lmonths = mean(ln(mm),ln(mf),ln(fm),   average log months.
            ln(ff)).
            x = birth - size.                    align births.
/plot       yvar = lmonths.  xvar = x.
/end
2 2 39.8 39.5 39.4 39.3
3 2 31.0 31.5 31.4 31.1                          15 lines of data as before
  . . .
```

Output B.3 shows the plot, using letters A, . . . , E to identify the five family sizes (except where an asterisk denotes multiple points). The systematically shorter interval between births 1 and 2 is identifiable. Output B.3 shows also the regression equations which the program prints but which for this problem are not required. Table B.3, *App. Stat.*, summarizing the birth interval pattern, is best obtained by hand.

Output B.1

Range and average of log months, and antilog of average. BMDP1D

C A S E NO. LABEL	1 size	7 range	8 lmonths	9 months
1	2	.0126	3.676	39.500
2	3	.0160	3.442	31.249
3	3	.0208	3.768	43.299
4	4	.0322	3.330	27.948
5	4	.0231	3.540	34.474
6	4	.0344	3.770	43.371
7	5	.0118	3.239	25.500
8	5	.0133	3.404	30.075
9	5	.0292	3.527	34.023
10	5	.0192	3.731	41.724
11	6	.0206	3.189	24.274
12	6	.00725	3.318	27.600
13	6	.0330	3.410	30.273
14	6	.0237	3.526	33.973
15	6	.0272	3.700	40.448

Output B.2

Mean of variables given in Output B.1. BMDP1D

VARIABLE NO. NAME	TOTAL FREQUENCY	MEAN	STANDARD DEVIATION
1 size	15	4.667	1.291
7 range	15	0.022	0.008
8 lmonths	15	3.505	0.193
9 months	15	33.849	6.475

Output B.3

Average of log months versus order, x, in family. BMDP6D

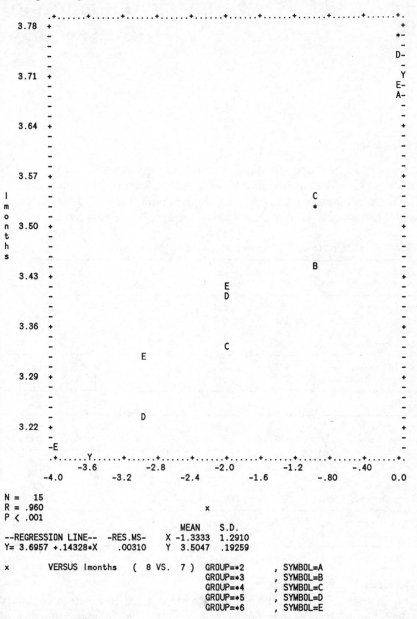

N = 15
R = .960   x
P < .001
                                  MEAN   S.D.
--REGRESSION LINE-- -RES.MS-   X -1.3333 1.2910
Y= 3.6957 +.14328*X   .00310   Y 3.5047   .19259

x   VERSUS lmonths (8 VS. 7) GROUP=*2 , SYMBOL=A
                                      GROUP=*3 , SYMBOL=B
                                      GROUP=*4 , SYMBOL=C
                                      GROUP=*5 , SYMBOL=D
                                      GROUP=*6 , SYMBOL=E

Note: a line drawn to join the two Ys on the frame of the plot would represent the regression of Y on X.

Example C   Statistical aspects of literary style

Description of data

As part of an investigation of the authorship of works attributed to St Paul, Morton (1965) found the numbers of sentences having zero, one, two, . . . occurrences of 'kai' ($\equiv$ and) in some of the Pauline works. Table C.1 gives a slightly condensed form of the resulting frequency distributions.

Table C.1. Frequency of occurrences of 'kai' in 10 Pauline works

Number of sentences with	Romans (1–15)	1st Corinth.	2nd Corinth.	Galat.	Philip.
0 kai	386	424	192	128	42
1 kai	141	152	86	48	29
2 kai's	34	35	28	5	19
3 or more kai's	17	16	13	6	12
No. of sentences	578	627	319	187	102
Total number of kai's	282	281	185	82	107

Number of sentences with	Colos.	1st Thessal.	1st Timothy	2nd Timothy	Hebrews
0 kai	23	34	49	45	155
1 kai	32	23	38	28	94
2 kai's	17	8	9	11	37
3 or more kai's	9	16	10	4	24
No. of sentences	81	81	106	88	310
Total number of kai's	99	99	91	68	253

The analysis

The analysis for this problem is nonstandard apart from the preliminary calculation of means and standard deviations. The summary statistics of Table C.2, *App. Stat.*, i.e. mean number of kai's, modified mean (taking 3 or more kai's equal to 3), modified standard deviation (using the modified data),

22

ratio of variance to mean, and standard error of the mean, being simply transformations down each column of data are easily obtained using BMDP1D.

The consistency within any proposed group of k means $\bar{Y}_1, \ldots, \bar{Y}_k$ with standard errors $\sqrt{v_1}, \ldots, \sqrt{v_k}$ is tested by the χ^2 statistic with $k-1$ degrees of freedom

$$\chi^2 = \Sigma(\bar{Y}_j - \bar{Y}.)^2/v_j \qquad \text{(C.1, App. Stat.)}$$

$$= \Sigma \bar{Y}_j^2/v_j - (\Sigma \bar{Y}_j/v_j)^2(\Sigma 1/v_j)^{-1},$$

where $\bar{Y}. = (\Sigma \bar{Y}_j/v_j)(\Sigma 1/v_j)^{-1}$.

This χ^2 statistic can be computed also by BMDP1D, using the **retain** instruction to retain values formed in the **transform** paragraph for use with the next line of data, thereby permitting calculation of $\Sigma 1/v_j$, $\Sigma \bar{Y}_j/v_j$ and $\Sigma \bar{Y}_j^2/v_j$, and thence the value of χ^2. In this way the χ^2 statistic can readily be obtained for any grouping of the literary works.

Program

BMDP1D is used. Since this operates on each line of data as it is read, the data must be arranged so that each line corresponds to one of the Pauline works.

Instructions for computing and printing the summary statistics of Table C.2, *App. Stat.* are as follows:

```
/problem     title = 'example c. literary style.
             statistics. bmdp1d.'.
/input       variables = 7.
             format = free.
/variable    names = pwk, f0, f1, f2, f3, ns, nk.    names of input variables
             add = new.
             use = pwk, m, yj, s, ratio, seyj.
/transform   m = nk/ns.                               mean no. kai's
             yj = (f1 + 2*f2 + 3*f3)/ns.              modified mean Ȳⱼ
             var = ((f1 + 4*f2 + 9*f3) -
             ns*yj**2)/(ns - 1).
             s = sqrt(var).                           modified s.d. (s)
             ratio = var/yj.                          s²/Ȳⱼ
             vj = var/ns.
             seyj = sqrt(vj).                         s.e.(Ȳⱼ)
/print       data.
/end
1 386 141 34 17 578 282                              10 lines of data,
2 424 152 35 16 627 281                                one line for each P-work
   . . .
```

The output lists all the variables. Only those corresponding to the summary statistics of Table C.2, *App. Stat.* are shown in Output C.1.

The following instructions, also for BMDP1D, compute the χ^2 statistic. To compute χ^2 for a subset of the Pauline works we use **omit** in the **transform** paragraph. Thus to compute χ^2 for the group of Pauline works 2, 3 and 4 the instructions are:

```
/problem      title = 'example c. literary style.
              bmdp1d.'.
/input        variables = 7.
              format = free.
/variable     names = pwk, f0, f1, f2, f3, ns, nk.
              add = new. retain.
              use = pwk, chi.
/transform    omit = 1, 5 to 10.                      omit P-works 1, 5, etc.
              m = nk/ns.
              yj = (f1 + 2*f2 + 3*f3)/ns.
              var = ((f1 + 4*f2 + 9*f3) -
              ns*yj**2)/(ns - 1).
              vj = var/ns.
              if (kase eq 2) then (sum0 = 1/vj.
              sum1 = yj/vj.  sum2 = (yj**2)/vj.
              chi = 0.).
              if (kase gt 2) then (sum0 = sum0+1/vj.  sum0=Σ1/vⱼ, sum1=ΣȲⱼ/vⱼ
              sum1 = sum1+yj/vj.  sum2 = sum2 +       sum2=ΣȲⱼ²/vⱼ, chi=χ²
              (yj**2)/vj.  chi = sum2 -
              (sum1**2)/sum0.).
/print        data.
/end
1 386 141 34 17 578 282                               10 lines of data as before
2 424 152 35 16 627 281
      . . .
```

The equations in the right margin read:
$$\text{sum0} = \Sigma 1/v_j, \quad \text{sum1} = \Sigma \bar{Y}_j/v_j$$
$$\text{sum2} = \Sigma \bar{Y}_j^2/v_j, \quad \text{chi} = \chi^2$$

The value of χ^2 is printed successively for each line of data. The final line in Output C.2 shows 7.654 for the value of the χ^2 statistic for the group of works 2, 3 and 4; the degrees of freedom will be 2. In this way the χ^2 statistics for each of the groups listed (and others) in Table C.3, *App. Stat.* can be obtained. Note for a particular subgroup it will be necessary to amend not only the **omit** sentence but also both conditional statements, i.e. **if (kase eq —)** and **if kase gt —)**, where — is replaced by the lowest number of the Pauline work in the group.

Suggested further work

Fit a censored negative binomial distribution by the method of maximum likelihood to each literary work. Compare the resulting analysis with that given here.

Output C1

Summary statistics for 10 Pauline works. BMDP1D

DATA AFTER TRANSFORMATIONS

C A S E NO. LABEL	1 pwk	8 m	9 yj	11 s	12 ratio	14 seyj
1	1	.488	.450	.737	1.206	.0306
2	2	.448	.431	.715	1.186	.0285
3	3	.580	.567	.817	1.177	.0457
4	4	.439	.406	.700	1.205	.0512
5	5	1.049	1.010	1.039	1.069	.103
6	6	1.222	1.148	.963	.808	.107
7	7	1.222	1.074	1.149	1.228	.128
8	8	.858	.811	.947	1.106	.0920
9	9	.773	.705	.860	1.049	.0917
10	10	.816	.774	.939	1.138	.0533

pwk = Pauline work, m = mean, yj = modified mean, s = modified s.d., ratio = s^2/(modified mean), seyj = s.e. (modified mean).

Output C2

χ^2 statistic for group containing Pauline works 2, 3 and 4. BMDP1D

C A S E NO. LABEL	1 pwk	15 chi
2	2	0
3	3	6.434
4	4	7.654

Example D

Temperature distribution in a chemical reactor

Description of data*

A chemical reactor has 1250 sections and it is possible to calculate a theoretical temperature for each section. These have a distribution across sections with mean 452 °C and standard deviation 22 °C; the distribution is closely approximated by a normal distribution in the range 390–520 °C. For a variety of reasons, measured temperatures depart from the theoretical ones, the discrepancies being partly random and partly systematic. The temperature is measured in 20 sections and Table D.1 gives the measurements and the corresponding theoretical values. It is known further that the measured temperature in a section deviates from the 'true' temperature by a completely random error of measurement of zero mean and standard deviation 3 °C. Special interest attaches to the number of channels in the reactor with 'true' temperature above 490 °C.

Table D.1. Measured and theoretical temperatures in 20 sections of reactor

Measured temp (°C)	Theoretical temp (°C)	Measured temp (°C)	Theoretical temp (°C)
431	432	472	498
450	470	465	451
431	442	421	409
453	439	452	462
481	502	451	491
449	445	430	416
441	455	458	481
476	464	446	421
460	458	466	470
483	511	476	477

*Fictitious data based on a real investigation.

26

The analysis

The working assumption made in this problem is that over the reactor we can treat the true and measured temperatures T and T_{ME} as random with

$$T = \alpha + \beta(t_{TH} - t) + \varepsilon,$$
$$T_{ME} = T + \varepsilon',$$

(D.1, *App. Stat.*)

where t is the mean theoretical temperature in the measured channels, ε is a random term, and ε' is a measurement error independent of T.

Standard linear regression of measured temperature on theoretical temperature can be used to obtain an estimate $\hat{\beta}$ of the slope and hence of the mean of the true temperature, i.e. $\bar{T}_{ME} + \hat{\beta}(452 - \bar{t}_{TH})$ and standard deviation of this estimated mean, i.e. $\sqrt{(484\hat{\beta}^2 + s^2 - 9)}$, where s^2 is the residual mean square.

The proportion of values of T above 490 is estimated by hand assuming normality and approximate confidence limits calculated via large-sample theory as discussed in *App. Stat.*

Program

BMDP1R is used for linear regression of measured temperature on theoretical temperature. The instructions are:

```
/problem    title = 'example d. chemical reactor.
            bmdp1r'.
/input      variables = 2. format = free.
/variable   names= time, tth.
/regress    dependent = time.
/print      data.                           print data, residuals, predicted values
/plot       residual. normal.               residuals v. predicted values and
/end                                         normal probability plot
431 432
450 470                                      20 lines of data
...
```

The **print** and **plot** paragraphs are optional, but are useful for checking the adequacy of the regression. Output D.1 gives details of the estimated regression; Output D.2 shows the residual and normal plots. The program also lists the predicted values and residuals (not shown here).

Other program

BMDP2R may also be used.

Suggested further work

Repeat the regression taking logarithms (add a **transform** paragraph) and compare the results.

Output D.1

Regression of measured temperature on theoretical temperature. BMDP1R

VARIABLE	MEAN	STANDARD DEVIATION	COEFFICIENT OF VARIATION	MINIMUM	MAXIMUM
1 tme	454.6000	18.0391	0.03968	421.00000	483.00000
2 tth	459.7000	28.7092	0.06245	409.00000	511.00000

```
    REGRESSION TITLE IS
example d.  chemical reactor.  bmdp1r

    DEPENDENT VARIABLE. . . . . . . . . . . . . .      1 tme
    TOLERANCE . . . . . . . . . . . . . . . . .  0.0100
ALL DATA CONSIDERED AS A SINGLE GROUP

MULTIPLE R           0.8119           STD. ERROR OF EST.       10.8209
MULTIPLE R-SQUARE    0.6591

ANALYSIS OF VARIANCE
                     SUM OF SQUARES   DF   MEAN SQUARE    F RATIO   P(TAIL)
          REGRESSION     4075.1523     1    4075.1523     34.803    0.0000
          RESIDUAL       2107.6428    18     117.0913

                                            STD. REG
    VARIABLE     COEFFICIENT   STD. ERROR     COEFF        T   P(2 TAIL) TOLERANCE

INTERCEPT         220.09731
tth        2        0.51012      0.08647     0.812     5.899    0.0000    1.00000
```

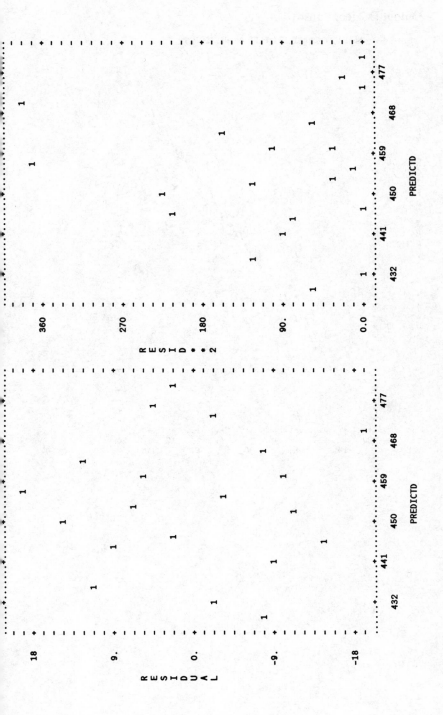

Output D.2 (continued)

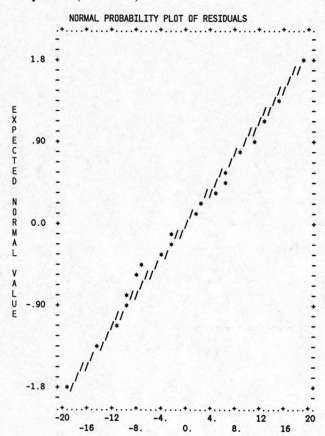

Example E

A 'before and after' study of blood pressure

Description of data

Table E.1 gives, for 15 patients with moderate essential hypertension, supine systolic and diastolic blood pressures immediately before and two hours after taking 25 mg of the drug captopril. The data were provided by Dr G. A. MacGregor, Charing Cross Hospital Medical School; for a report on the investigation and appreciable further summary data, see MacGregor, Markandu, Roulston and Jones (1979).

Table E.1. Blood pressures (mm Hg) before and after captopril

Patient no.	Systolic			Diastolic		
	before	*after*	*difference*	*before*	*after*	*difference*
1	210	201	−9	130	125	−5
2	169	165	−4	122	121	−1
3	187	166	−21	124	121	−3
4	160	157	−3	104	106	2
5	167	147	−20	112	101	−11
6	176	145	−31	101	85	−16
7	185	168	−17	121	98	−23
8	206	180	−26	124	105	−19
9	173	147	−26	115	103	−12
10	146	136	−10	102	98	−4
11	174	151	−23	98	90	−8
12	201	168	−33	119	98	−21
13	198	179	−19	106	110	4
14	148	129	−19	107	103	−4
15	154	131	−23	100	82	−18

The analysis

Main interest here lies simply in plotting the data. Treating the difference in blood pressure as the response measure, we plot the difference against the 'before' value for both systolic and diastolic pressures, and the systolic difference against the diastolic difference. The BMDP program automatically computes also the regression for each plot.

Program

BMDP6D prints scatter plots and computes the corresponding regressions. Writing **sys0, sys1, dya0, dya1** to denote the 'before' and 'after' systolic and dyastolic measurements, and **diffs, diffd** the differences 'after' minus 'before', the instructions for obtaining the three scatter plots and corresponding regressions are:

```
/problem     title = 'example e. before/after.
             bmdp6d'.
/input       variables = 4. format = free.
/variable    names = sys0, sys1, dya0, dya1.
             add = new.
/transform   diffs = sys1 - sys0.              'after minus 'before'
             diffd = dya1 - dya0.
/plot        yvar = diffs, diffd, diffs.
             xvar = sys0, dya0, diffd.         three plots requested
             statistics. size = 50, 30.        print means and regressions
/end
210 201 130 125                               15 lines of data
169 165 122 121
    . . .
```

Note we need read in only the 'before' and 'after' measurements, leaving the program to compute the differences. Output E.1 shows the plot for systolic difference against dyastolic difference, i.e. Fig. E.1, *App. Stat.*; it gives also the mean and s.d. for the systolic and dyastolic differences (quoted in Table E.2, *App. Stat.*) and the equations of the two regression lines. Output for **diffs** against **sys0**, and **diffd** against **dya0** is similar.

Suggested further work

(i) Repeat the analysis using log blood pressures. How would you decide (perhaps given more extensive data) which analysis is to be preferred?

(ii) Suppose it is suggested that log(diastolic × systolic) and log(diastolic/systolic) are good indices of overall blood pressure and blood pressure profile, respectively. How would you proceed?

A 'before and after' study of blood pressure 33

Output E.1

Systolic difference ('after' minus 'before') versus dyastolic difference.
BMDP6D

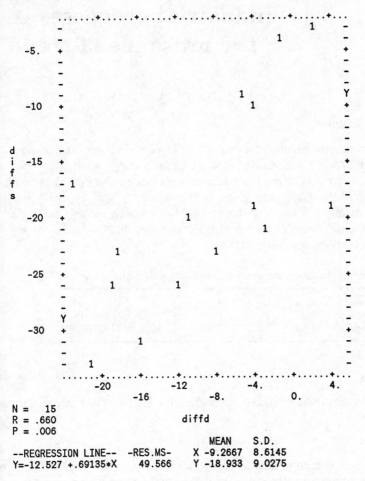

N = 15
R = .660
P = .006

--REGRESSION LINE--   -RES.MS-   MEAN   S.D.
Y=-12.527 +.69135*X   49.566   X -9.2667 8.6145
                               Y -18.933 9.0275

Example F

Comparison of industrial processes in the presence of trend

Description of data*

In a plant-scale experiment on the production of a certain chemical, a batch of intermediate product was divided into six equal portions which were then processed on successive days by two different methods, P_1 and P_2. The order of treatment and the yields are given in Table F.1. It was expected that superposed on any process effect there would be a smooth, roughly parabolic trend. Experience of similar experiments showed that the standard deviation of a single observation was about 0.1.

Table F.1. Treatment and yields in plant-scale experiment

Day	1	2	3	4	5	6
Process	P_1	P_2	P_2	P_1	P_1	P_2
Yield	5.84	5.73	7.30	10.46	9.71	5.91

The analysis

We could use the multiple regression program BMDP1R to fit a model of the form

$$E(Y) = \mu \pm \tau + \beta t + \gamma t^2$$

where $t = 1, \ldots, 6$ denotes day, and $+/-$ correspond to Process 1/2. There can however be an advantage to using orthogonal polynomials to represent the quadrate trend, particularly if the time delay to maximum yield, and its standard error, are to be estimated. In this case we write the model as

$$E(Y) = \mu \pm \tau + \beta_1(2t - 7) + \beta_2(28 - 21t + 3t^2)/2$$
$$= \mu \pm \tau + \beta_1 \varphi_1 + \beta_2 \varphi_2,$$

where φ_1 and φ_2 denote linear and quadratic orthogonal polynomials; φ_1 and φ_2 are computed in the **transform** paragraph of the program, although with so

* Fictitious data based on a real investigation.

34

few lines of data we could equally well have chosen to read in their values as data.

An approximate variance for the estimated time delay corresponding to maximum yield is given by substituting estimated values into the expression

$$\tfrac{4}{9}\{\beta_2^2 \, \mathrm{var}(\hat{\beta}_1) + \beta_1^2 \, \mathrm{var}(\beta_2)\}/\beta_2^4 \qquad \text{(F.4, } App. \ Stat.\text{)}$$

Program

BMDP1R is used. The instructions are as follows:

```
/problem     title = 'example f.  industrial
             process with trend.  bmdp1r.'.
/input       variables = 3.  format = free.
/variable    names =  day, process, yield.
             add = new.
/transform   phi1 = 2*day - 7.                    compute φ₁
             phi2 = (28 - 21*day + 3*day**2)/2.   compute φ₂
             process = 3 - 2*process.             recode process as ±1
/regress     depend = yield.
             independ = phi1, phi2, process.
/print       data.                                print data
/plot        variable = day.                      plot predicted values
/end                                              6 lines of data
1 1 5.84
2 2 5.73
  . . .
```

Output F.1 shows details of the regression, in particular the residual mean square of 0.0436 (for comparison with $\sigma^2 = 0.01$) the estimated coefficients and standard errors, and the predicted values and residuals. Output F.2 shows the observed values and the residuals plotted against 'day'; the asterisks indicate an overlap of observed and predicted values, the residuals being small in relation to the observed values.

The estimated time delay $\hat{t}$ corresponding to maximum yield is given by $\hat{t} = 3.5 - (2\hat{\beta}_1)/(3\hat{\beta}_2)$ and $\mathrm{var}(\hat{t})$ is estimated by (F.4, *App. Stat.*) given above.

Other program

BMDP2R may be used, with the **set** option being used to permit non-stepwise entry of variables.

Suggested further work

Repeat the analysis taking the model in the forms
(a) $\mu \pm \tau + \beta t + \gamma t^2$.
(b) $\mu + \beta t + \gamma t^2$ (Process 1), $\mu + \Delta + \beta t + \gamma t^2$ (Process 2) and make a careful comparison of the answers obtained from (a) and (b) with those obtained previously.

Output F.1

Regression of yield on process and time trend. BMDP1R

```
DEPENDENT VARIABLE. . . . . . . . . . . . .   3 yield
TOLERANCE . . . . . . . . . . . . . . . . .   0.0100
ALL DATA CONSIDERED AS A SINGLE GROUP

MULTIPLE R            0.9990        STD. ERROR OF EST.        0.1477
MULTIPLE R-SQUARE     0.9980
```

ANALYSIS OF VARIANCE

	SUM OF SQUARES	DF	MEAN SQUARE	F RATIO	P(TAIL)
REGRESSION	22.0582	3	7.3527	337.004	0.0030
RESIDUAL	0.0436	2	0.0218		

VARIABLE		COEFFICIENT	STD. ERROR	STD. REG COEFF	T	P(2 TAIL)	TOLERANCE
INTERCEPT		7.49167					
process	2	1.26394	0.06059	0.659	20.860	0.0023	0.99048
phi1	4	0.25683	0.01774	0.457	14.478	0.0047	0.99048
phi2	5	-0.33012	0.01612	-0.644	-20.483	0.0024	1.00000

```
LIST OF PREDICTED VALUES, RESIDUALS, AND VARIABLES
   NOTE - NEGATIVE CASE NUMBER DENOTES A CASE WITH MISSING VALUES.
          THE NUMBER OF STANDARD DEVIATIONS FROM THE MEAN IS DENOTED BY UP TO 3 ASTERISKS TO THE RIGHT
          OF EACH RESIDUAL OR VARIABLE.

MISSING VALUES AND VALUES OUT OF RANGE ARE DENOTED BY VALUES
GREATER THAN OR EQUAL TO  0.2127E+38 IN ABSOLUTE VALUE.
```

CASE LABEL	NO.	RESIDUAL	PREDICTED VALUE	VARIABLES 1 day		2 process		njyield		4 phi1		5 phi2	
	1	0.1912E-01	5.821	1.000	*	1.000		5.840		-5.000	*	5.000	*
	2	-0.5736E-01	5.787	2.000		-1.000		5.730		-3.000		-1.000	
	3	0.8626E-02	7.291	3.000		-1.000		7.300		-1.000		-4.000	
	4	0.1271	10.33	4.000		1.000		10.46	*	1.000		-4.000	
	5	-0.1462	9.856	5.000		1.000		9.710	*	3.000		-1.000	
	6	0.4874E-01	5.861	6.000	*	-1.000		5.910		5.000	*	5.000	*

36

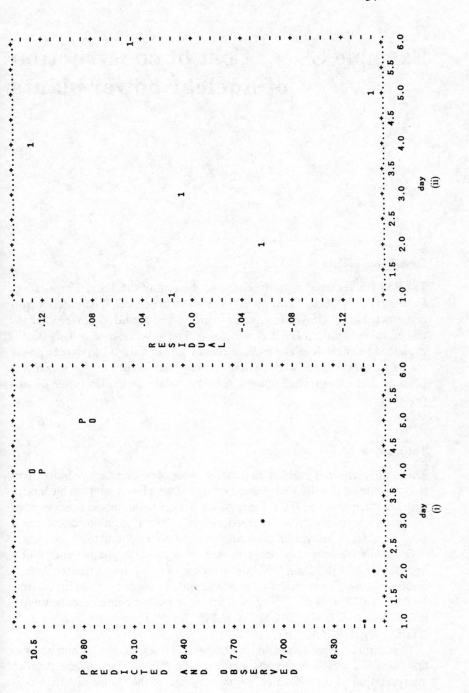

Example G   Cost of construction of nuclear power plants

Description of data

Table G.1 gives data, reproduced by permission of the Rand Corporation, from a report (Mooz, 1978) on 32 light water reactor (LWR) power plants constructed in the USA. It is required to predict the capital cost involved in the construction of further LWR power plants. The notation used in Table G.1 is explained in Table G.2. The final six lines of data in Table G.1 relate to power plants for which there were partial turnkey guarantees and for which it is possible that some manufacturers' subsidies might be hidden in the quoted capital costs.

The analysis

This is an interesting problem in that a number of models may be found to fit the data almost equally well. Table G.4 *App. Stat.* gives a six-variable model, with log C regressed on PT, CT, log N, log S, D, NE, this model being selected by successive elimination. Interactions with PT, a variable identifying a subgroup of the observations, are examined but none is found to be significant.

Other models, not considered in *App. Stat.*, include the possibility of a $(\log N)^2$ term. Also, Ling (1984) in a review of *App. Stat.* suggests there is evidence for a different model when $N > 12$ and so defines an indicator variable $I = 0$ for $N \leqslant 12$, $I = 1$ for $N > 12$, and fits three models containing the sets: (i) CT, T, S, D, NE, $D \times N$; (ii) CT, I, S, D, NE, $D \times N$, N; and (iii) CT, I, log S, D, NE, $D \times$ log N, log N.

The adequacy of all competing models should be examined, the aim being to find a model which is a good predictor of capital cost. Where detailed interpretation of the estimated parameters and of the choice of variables is intrinsically important, especial care is necessary.

38

Table G.1. Data on 32 LWR power plants in the USA

C	D	T_1	T_2	S	PR	NE	CT	BW	N	PT
460.05	68.58	14	46	687	0	1	0	0	14	0
452.99	67.33	10	73	1065	0	0	1	0	1	0
443,22	67.33	10	85	1065	1	0	1	0	1	0
652.32	68.00	11	67	1065	0	1	1	0	12	0
642.23	68.00	11	78	1065	1	1	1	0	12	0
345.39	67.92	13	51	514	0	1	1	0	3	0
272.37	68.17	12	50	822	0	0	0	0	5	0
317.21	68.42	14	59	457	0	0	0	0	1	0
457.12	68.42	15	55	822	1	0	0	0	5	0
690.19	68.33	12	71	792	0	1	1	1	2	0
350.63	68.58	12	64	560	0	0	0	0	3	0
402.59	68.75	13	47	790	0	1	0	0	6	0
412.18	68.42	15	62	530	0	0	1	0	2	0
495.58	68.92	17	52	1050	0	0	0	0	7	0
394.36	68.92	13	65	850	0	0	0	1	16	0
423.32	68.42	11	67	778	0	0	0	0	3	0
712.27	69.50	18	60	845	0	1	0	0	17	0
289.66	68.42	15	76	530	1	0	1	0	2	0
881.24	69.17	15	67	1090	0	0	0	0	1	0
490.88	68.92	16	59	1050	1	0	0	0	8	0
567.79	68.75	11	70	913	0	0	1	1	15	0
665.99	70.92	22	57	828	1	1	0	0	20	0
621.45	69.67	16	59	786	0	0	1	0	18	0
608.80	70.08	19	58	821	1	0	0	0	3	0
473.64	70.42	19	44	538	0	0	1	0	19	0
697.14	71.08	20	57	1130	0	0	1	0	21	0
207.51	67.25	13	63	745	0	0	0	0	8	1
288.48	67.17	9	48	821	0	0	1	0	7	1
284.88	67.83	12	63	886	0	0	0	1	11	1
280.36	67.83	12	71	886	1	0	0	1	11	1
217.38	67.25	13	72	745	1	0	0	0	8	1
270.71	67.83	7	80	886	1	0	0	1	11	1

Table G.2. Notation for data of Table G.1

C Cost in dollars $\times 10^{-6}$, adjusted to 1976 base
D Date construction permit issued
T_1 Time between application for and issue of permit
T_2 Time between issue of operating license and construction permit
S Power plant net capacity (MWe)
PR Prior existence of an LWR on same site ($=1$)
NE Plant constructed in north-east region of USA ($=1$)
CT Use of cooling tower ($=1$)
BW Nuclear steam supply system manufactured by Babcock-Wilcox ($=1$)
N Cumulative number of power plants constructed by each architect-engineer
PT Partial turnkey plant ($=1$)

A specified model may be fitted using BMDP1R or BMDP2R (a stepwise program which by use of the **set** option permits non-stepwise entry). If used for stepwise selection, BMDP2R must, as with any automatic procedure, be used with much caution to arrive at an appropriate selection; it here leads to the six-variable model of Table G.4, *App. Stat.* BMDP9R compares all possible subsets of regressions, often useful after preliminary reduction of the problem. All three programs can list or plot residuals for scrutiny.

Either BMDP1R or BMDP2R may be used to fit the sequence of models. Instructions are given illustrating both programs.

Program

BMDP1R can be used to reduce the number of variables from the full set of 10 by successively fitting specified models, as done in Table G.3, *App. Stat.* Note we take logarithms of C, N, S, T_1 and T_2. The instructions for fitting the model containing the full 10 variables are:

```
/problem      title = 'example g.  power plants.
              bmdp1r.'.
/input        variables = 11.  format = free.
/variable     names  =  c, d, t1, t2, s, pr, ne,
              ct, bw, n, pt.  add = new.
/transform    lc = ln(c).
              ln = ln(n).                           take logs of
              ls = ln(s).                           C, N, S, T₁, T₂
              lt1 = ln(t1).
              lt2 = ln(t2).
/regress      depend = lc.
              independent = d, pr to bw, pt,
              ln to lt2.
/print        data.                                 print fitted values, residuals
/plot         variable = ln, d, ls.  residual.      residual plots
              normal.
/end
460.05 68.58 14 46 687 0 1 0 0 14 0                  32 lines of data
452.99 67.33 10 73 1065 0 0 1 0 1 0
        . . .
```

The **print** and **plot** paragraphs are optional and can be deleted, or amended to give plots against other variables. The **variable** sentence of the **plot** paragraph as shown requests plots of residuals against only LN, D and LS. Output G.1 shows some of the output for the full ten variable-regression; in particular it gives the residual s.s. 0.5680 as in Table G.3, *App. Stat.*, and the estimates and s.e.s as in Table G.4, *App. Stat.* The plots are not shown.

For successive fitting of alternative models, the list of variables in the **independent** sentence of the **regress** paragraph must be amended to list only these variables to be included. Several models may be fitted in the same computer run by adding further **regress** paragraphs.

If interactions between PT and other variables, as in Table G.5, *App. Stat.*, are to be included the **transform** paragraph will need an additional sentence such as

ptxct = pt∗ct.

and the name **ptxct** should be added to those in the **independent** sentence of the **regress** paragraph. Additional sentences for other interactions, or powers of variables, can be added at the same time.

If a term $(\log N)^2$ is to be included, as well as $\log N$, then

tolerance = 0.001.

should be included in the **regress** paragraph; the default value is 0.01 and the program excludes any variables whose correlation with other variables exceeds $\sqrt{(1 - \text{tolerance})}$. It will also be beneficial to work with $\log N$ minus its average value (1.76894, printed in Output G.1) rather than simply $\log N$. (Any convenient constant close to the average will do almost as well.) The **transform** paragraph will need additional statements:

ln = ln − 1.76894.
lnsq = ln∗∗2.

where **lnsq** denotes $(\log N)^2$.

BMDP2R is a stepwise procedure. As noted above, considerable caution is needed in its use. In its simplest form it is a forwards procedure, building up the model. It can be used also for backward stepping, eliminating variables, or, by specifying **level** values, the order in which variables enter the model can be controlled. Also, use of the **set** option permits non-stepwise entry. We here use BMDP2R to illustrate (i) fitting the full regression on 10 variables (as done above with BMDP1R) and (ii) a forward stepwise selection. The instructions include two **regress** paragraphs:

```
/problem    title = 'example g.  power plants.
            bmdp2r.'.
/input      variables = 11.  format = free.
/variable   names  =   c, d, t1, t2, s, pr, ne,
            ct, bw, n, pt.  add = new.
/transform  lc = ln(c).
            ln = ln(n).
            ls = ln(s).
            lt1 = ln(t1).
            lt2 = ln(t2).
/regress    title = 'stepwise selection'.          forward selection
            dependent = lc.
            independent = d, pr to bw, pt,
            ln to lt2.
/regress    title = 'fitting all 10 variables'.    10-variable regression as with
            dependent = lc.                              BMDP1R
            independent = a.  setn = a.            set option
            a = d, pr to bw, pt, ln to lt2.
/plot       variable = ln, d, ls.  residual.
            normal.
/end
460.05 68.58 14 46 687 0 1 0 0 14 0               32 lines of data as before
452.99 67.33 10 73 1065 0 0 1 0 1 0
            . . .
```

Forward selection leads to the six-variable model given in *App. Stat.* Outputs G.2–G.5 show some of the output at the final step. The estimates and s.e.s shown correspond to those in Table G.4, *App. Stat.* Plots of the residuals against *D*, against predicted values and the normal, probability plot, corresponding to Figs. G.1, G.3 and G.4, *App. Stat.*, are also shown.

BMDP9R compares regressions based on all possible subsets of explanatory variables and can be used for examination of further reduced sets of the variables. It prints the value of R^2 and Mallows' C_p for all subsets, and details of the estimated regression for the five selected 'best' subsets. The criterion for selection can be R^2, adjusted R^2, or C_p, chosen by insertion of a **method** statement in the **regression** paragraph; if omitted, adjusted R^2 is used. Instructions for selecting subsets from seven variables (the above six plus **lnsq**) are:

```
/problem      title = 'example g.  power plants.
              bmdp9r.'.
/input        variables = 11. format = free.
/variable     names  =  c, d, t1, t2, s, pr, ne, ct,
              bw, n, pt.  add = new.
/transform    lc = ln(c).
              ln = ln(n).
              ls = ln(s).
              lt1 = ln(t1).
              lt2 = ln(t2).
              ln = ln - 1.76894.
              lnsq = ln**2.
/regress      dependent = lc.
              independent = pt, ct, ln, ls, d,
              ne, lnsq.
/end
460.05 68.58 14 46 687 0 1 0 0 14 0          32 lines of data as before
452.99 67.33 10 73 1065 0 0 1 0 1 0
```

. . .

Output G.6 shows some of the results for subsets of size 5 to 7 variables. The 'best' subset selected by the program is the first set of 6 in Output G.6 with adjusted $R^2 = 0.822532$, i.e. the previously selected set. The full output (not shown here) gives also the residual m.s. from this model and s.e.s of the estimated coefficients.

Additional diagnostic information (including Cook's distance) for the 'best' subset may be obtained by inserting

/print   residuals.

after the **regress** paragraph, and

/plot   normal.

will give a normal probability plot.

Suggested further work

(i) Examine the adequacy of other possible subsets indicated by Output G.6, or of any other models felt worth considering.

(ii) Comment critically on the limitations of R^2 as a criterion for the comparison of fits to different sets of data.

Output G.1

Ten-variable regression. BMDP1R

VARIABLE	MEAN	STANDARD DEVIATION	COEFFICIENT OF VARIATION	MINIMUM	MAXIMUM
1 c	461.5603	170.1207	0.36858	207.50999	881.23999
2 d	68.5813	1.0153	0.01480	67.17000	71.08000
3 t1	13.7500	3.3697	0.24507	7.00000	22.00000
4 t2	62.3750	10.3946	0.16665	44.00000	85.00000
5 s	825.3750	189.3592	0.22942	457.00000	1130.00000
6 pr	0.3125	0.4709	1.50697	0.00000	1.00000
7 ne	0.2500	0.4399	1.75977	0.00000	1.00000
8 ct	0.4063	0.4990	1.22829	0.00000	1.00000
9 bw	0.1875	0.3966	2.11497	0.00000	1.00000
10 n	8.5313	6.3296	0.74193	1.00000	21.00000
11 pt	0.1875	0.3966	2.11497	0.00000	1.00000
12 lc	6.0672	0.3779	0.06229	5.33518	6.78133
13 ln	1.7689	0.9829	0.55567	0.00000	3.04452
14 ls	6.6876	0.2487	0.03718	6.12468	7.02997
15 lt1	2.5920	0.2465	0.09510	1.94591	3.09104
16 lt2	4.1195	0.1689	0.04101	3.78419	4.44265

REGRESSION TITLE IS
example g. power plants. bmdp1r

DEPENDENT VARIABLE. . . . .   12 lc
TOLERANCE 0.0100
ALL DATA CONSIDERED AS A SINGLE GROUP

MULTIPLE R   0.9337   STD. ERROR OF EST.   0.1645
MULTIPLE R-SQUARE   0.8717

ANALYSIS OF VARIANCE

	SUM OF SQUARES	DF	MEAN SQUARE	F RATIO	P(TAIL)
REGRESSION	3.8600	10	0.3860	14.271	0.0000
RESIDUAL	0.5680	21	0.0270		

VARIABLE		COEFFICIENT	STD. ERROR	STD. REG COEFF	T	P(2 TAIL)	TOLERANCE
INTERCEPT		-14.24198					
d	2	0.2092	0.0653	0.56	3.21	0.00	0.1988
pr	6	-0.0924	0.0773	-0.12	-1.19	0.25	0.6584
ne	7	0.2581	0.0769	0.30	3.35	0.00	0.7617
ct	8	0.1204	0.0663	0.16	1.82	0.08	0.7967
bw	9	0.0330	0.1011	0.03	0.33	0.75	0.5426
pt	11	-0.2243	0.1225	-0.24	-1.83	0.08	0.3700
ln	13	-0.0802	0.0460	-0.21	-1.74	0.10	0.4275
ls	14	0.6937	0.1361	0.46	5.10	0.00	0.7623
lt1	15	0.0919	0.2440	0.06	0.38	0.71	0.2413
lt2	16	0.2855	0.2729	0.13	1.05	0.31	0.4105

Output G.2

Six-variable model. Estimates of coefficients and s.e.s. BMDP2R

```
STEP NO.    6
---------------
VARIABLE ENTERED    13  In

MULTIPLE R              0.9257
MULTIPLE R-SQUARE      0.8569
ADJUSTED R-SQUARE      0.8225

STD. ERROR OF EST.     0.1592

ANALYSIS OF VARIANCE
                      SUM OF SQUARES      DF    MEAN SQUARE        F RATIO
          REGRESSION    3.7943258          6   0.6323876           24.95
          RESIDUAL      0.63374138        25   0.2534965E-01
```

VARIABLES IN EQUATION FOR Ic

VARIABLE		COEFFICIENT	STD. ERROR OF COEFF	STD REG COEFF	TOLERANCE	F TO REMOVE
(Y-INTERCEPT		-13.26033)				
d	2	0.21241	0.0433	0.571	0.42393	24.11
ne	7	0.24902	0.0741	0.290	0.76868	11.28
ct	8	0.14039	0.0604	0.185	0.89948	5.40
pt	11	-0.22610	0.1135	-0.237	0.40330	3.96
In	13	-0.08758	0.0415	-0.228	0.49202	4.46
Is	14	0.72341	0.1188	0.476	0.93662	37.07

Output G.3

Six-variable model. Residuals versus predicted values. BMDP2R

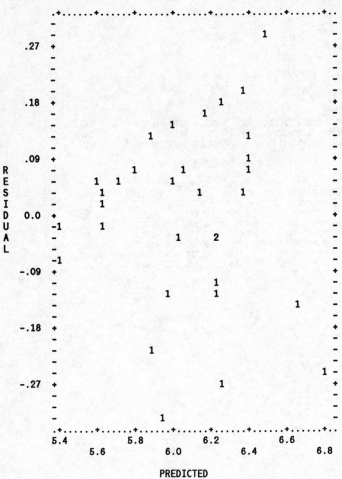

PREDICTED

Output G.4

Six-variable model. Residuals versus *D*. BMDP2R

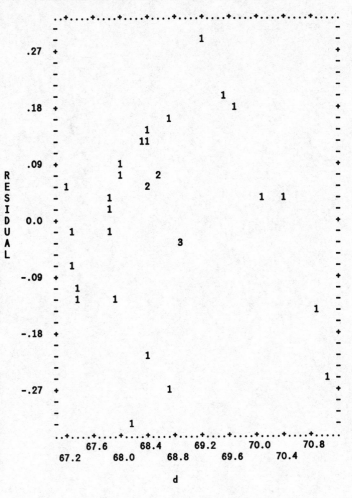

Output G.5

Six-variable model. Normal probability plot. BMDP2R

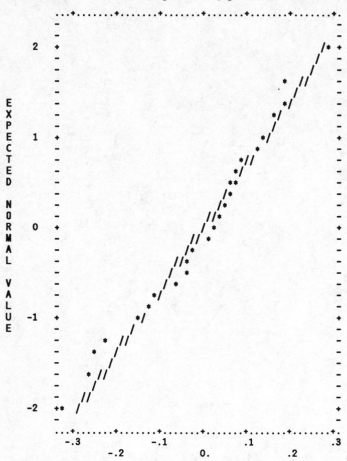

Output G.6

Best subsets regressions on PT, CT, LN, LS, *D*, NE, LNSQ. BMDP9R

SUBSETS WITH   5 VARIABLES

R-SQUARED	ADJUSTED R-SQUARED	cp	VARIABLE	COEFFICIENT	T-STATISTIC
0.835111	0.803402	8.32	11 pt	-0.342553	-3.46
			14 ls	0.647243	5.32
			2 d	0.144922	4.20
			7 ne	0.229141	3.12
			17 lnsq	0.0744442	2.17
			INTERCEPT	-8.26298	

Output G.6 (continued)

0.834182	0.802294	8.48	VARIABLE	COEFFICIENT	T-STATISTIC		
			8 ct	0.173854	2.84		
			13 ln	-0.137328	-3.93		
			14 ls	0.744601	5.96		
			2 d	0.273447	8.49		
			7 ne	0.311360	4.39		
			INTERCEPT	-17.8142			
0.831356	0.798924	8.96	pt	ct	ls	d	ne
0.825977	0.792510	9.89	pt	ln	ls	d	ne
0.796046	0.756825	15.03	ln	ls	d	ne	lnsq
0.789371	0.748865	16.17	ct	ls	d	ne	lnsq

SUBSETS WITH   6 VARIABLES

R-SQUARED	ADJUSTED R-SQUARED	cp					
0.856881	0.822532	6.58	VARIABLE	COEFFICIENT	T-STATISTIC		
			11 pt	-0.226104	-1.99		
			8 ct	0.140393	2.32		
			13 ln	-0.0875763	-2.11		
			14 ls	0.723407	6.09		
			2 d	0.212415	4.91		
			7 ne	0.249025	3.36		
			INTERCEPT	-13.4152			
0.850765	0.814949	7.63	VARIABLE	COEFFICIENT	T-STATISTIC		
			11 pt	-0.313604	-3.21		
			8 ct	0.102248	1.62		
			14 ls	0.650020	5.51		
			2 d	0.155169	4.55		
			7 ne	0.218169	3.05		
			17 lnsq	0.0616598	1.80		
			INTERCEPT	-9.01655			
0.838829	0.800147	9.68	pt lnsq	ln	ls	d	ne
0.836138	0.796812	10.14	ct lnsq	ln	ls	d	ne
0.795478	0.746393	17.12	pt lnsq	ct	ln	ls	d
0.759292	0.701522	23.34	pt lnsq	ct	ln	ls	ne
0.671767	0.592991	38.37	pt lnsq	ct	ln	d	ne

SUBSETS WITH   7 VARIABLES

R-SQUARED	ADJUSTED R-SQUARED	cp			
0.860248	0.819487	8.00	VARIABLE	COEFFICIENT	T-STATISTIC
			11 pt	-0.233972	-2.03
			8 ct	0.123987	1.92
			13 ln	-0.0652327	-1.28
			14 ls	0.701393	5.69
			2 d	0.198021	4.16
			7 ne	0.249375	3.34
			17 lnsq	0.0313855	0.76
			INTERCEPT	-12.3022	

Example H

Effect of process and purity index on fault occurrence

Description of data*

Minor faults occur irregularly in an industrial process and, as an aid to their diagnosis, the following experiment was done. Batches of raw material were selected and each batch was divided into two equal sections: for each batch, one of the sections was processed by the standard method and the other by a slightly modified process, in which the temperature at one stage is reduced. Before processing, a purity index was measured for the whole batch of material. For the product from each section of material, it was recorded whether the minor faults did or did not occur. Results for 22 batches are given in Table H.1.

Table H.1. Occurrence of faults in 22 batches

Purity index	Standard process	Modified process	Purity index	Standard process	Modified process
7.2	NF	NF	6.5	NF	F
6.3	F	NF	4.9	F	F
8.5	F	NF	5.3	F	NF
7.1	NF	F	7.1	NF	F
8.2	F	NF	8.4	F	NF
4.6	F	NF	8.5	NF	F
8.5	NF	NF	6.6	F	NF
6.9	F	F	9.1	NF	NF
8.0	NF	NF	7.1	F	NF
8.0	F	NF	7.5	NF	F
9.1	NF	NF	8.3	NF	NF

F, Faults occur.   NG, No faults occur.

The analysis

Preliminary analysis of the occurrence of F and NF within the matched pairs, given the small number of observations, is best done by hand. Then a logistic

* Fictitious data based on a real investigation.

50

model taking account of purity index and process can be fitted either by BMDP3R using the method of iterated weighted least squares, or by BMDPLR, a stepwise logistic regression program based on the method of maximum likelihood. Both programs lead to the estimates and asymptotic standard errors quoted in Table H.2, *App. Stat.*

BMDP3R will also plot fitted probabilities against purity index, corresponding to Fig. H.1, *App. Stat.* When using BMDP3R it is necessary to specify the model explicitly, which we write here as

$$\text{pr(fault } |x_i) = \frac{\exp(\alpha + \beta x_i \pm \Delta)}{1 + \exp(\alpha + \beta x_i \pm \Delta)}$$

rather than in the approximately orthogonal form with $(x_i - \bar{x})$ as in *App. Stat.*; it avoids feeding in the value of $\bar{x}$, the mean of the x_i, although convergence may be slower. The $+/-$ signs refer to standard/modified process, respectively, and x_i denotes the purity index of the ith batch.

Instructions are given for using both BMDP3R and BMDPLR.

Programs

BMDP3R is used to fit the above model by the method of iterated weighted least squares. The **fun** paragraph evaluates:

(i) pr (fault/x_i), denoted by **f**;
(ii) derivatives of **f** with respect to α, β, Δ, denoted by **df1**, **df2**, **df3**;
(iii) weight equal to $1/(\mathbf{f}(1-\mathbf{f}))$, denoted by **w**.

Parameters α, β and Δ must be denoted by **p1**, **p2** and **p3** when used in the **fun** paragraph. In the **regress** paragraph the **convergence** and **halving** statements cause the program to iterate towards a solution with zero derivatives (rather than minimum residual s.s.); the **meansquare** statement is necessary to produce the usual asymptotic information theory standard errors (see Dixon *et al.*, 1985, for further details). The instructions for using BMDP3R are as follows:

```
/problem     title = 'example h.  purity index.
             bmdp3r'.
/input       variables = 3.      format = free.
/variable    names = purity, y, process.
             add = new.
/transform   w = 1.0.
/fun         numer = exp(p1 + p2*purity +
             p3*process).
             f = numer/(1 + numer).
             df1 = f*(1 - f).
             df2 = purity*df1.
             df3 = process*df1.
             w = 1/(f*(1 - f)).
```

$p1 = \alpha$, $p2 = \beta$, $p3 = \Delta$

$f = \text{pr(fault}|x_i)$
$\mathbf{df1} = \partial f/\partial \alpha$
$\mathbf{df2} = \partial f/\partial \beta$
$\mathbf{df3} = \partial f/\partial \Delta$
$\mathbf{w} = \text{weight}$

```
/regress      dependent = y.
              parameters = 3.                      no. of parameters = 3
              weight = w.
              meansquare = 1.0.
              convergence = -1.0.
              iterations = 10.
              halving = 0.
/plot         variable = purity.                   plot predicted prob. v. purity
/end
7.2 0 1                                            22 lines of data
7.2 0 -1                                           Code: Y = 0 (NF), 1 (F) in col. 2
                                                         Process = 1 (standard),
                                                         −1 (modified) in col. 3

              . . .
```

To fit a reduced model with $\Delta = 0$, or $\beta = 0$, or $\beta = \Delta = 0$, the **fun** and **regress** paragraphs should be amended in the obvious way. Note that the program takes zero as a default value for the initial approximation for each parameter to be estimated; other starting values should be specified if convenient approximations are available.

Output H.1 shows the estimated values of the parameters and s.e.s for the full model. Output H.2 plots the predicted proportions against purity index; the predicted proportions for the two processes are clearly distinguishable. The full output (not shown here) includes also a list of the data and predicted proportions.

Alternatively we can use BMDPLR. This is a stepwise logistic regression but varieties may be forced into the model by a **model** sentence in the **regress** paragraph. Output includes the maximum likelihood estimates of the parameters and s.e.s, the maximized log likelihood and predicted probabilities. Instructions for fitting the full model with three parameters using BMDPLR are:

```
/problem      title = 'example h.  purity index.
              bmdplr'.
/input        variables = 3.  format = free.
/variable     names = purity, y, process.
/group        codes(2) = 1, 0.
              names(2) = f, nf.                    labels for output
/regress      dependent = y.
              interval = purity.                   purity is a continuous variable
              model = purity, process.             include purity and process
/end
7.2 0 1                                             22 lines of data as before
7.2 0 -1


              . . .
```

Output H.3 shows the maximized log likelihood (as in Table H.2, *App. Stat.*), estimates of the parameters and s.e.s. Note BMDPLR estimates the probability of the first named category in the **group** paragraph, and hence we write **codes(2) = 1,0.** to make the results agree with those of BMDP3R.

Other programs

BMDPAR will compute maximum likelihood estimates; instructions required are similar to those for BMDP3R except that derivatives are not needed.

Suggested further work

(i) Examine the effect of changing the initial approximations.
(ii) Fit reduced models having $\Delta = 0$ or $\beta = 0$ or $\beta = \Delta = 0$ and interpret the results.
(iii) Run BMDPLR as a stepwise program and compare critically the results with those obtained above.

Output H.1

Estimated values of parameters and s.e.s. BMDP3R

PARAMETER	ESTIMATE	ASYMPTOTIC STANDARD DEVIATION	TOLERANCE	
P1	4.028554	2.092103	0.0251598160	($p1 = \alpha$, $p2 = \beta$, $p3 = \Delta$)
P2	-0.604212	0.283814	0.0250725130	
P3	0.432250	0.335847	0.9763159034	

Output H.2

Predicted proportions versus purity index for full model. BMDP3R

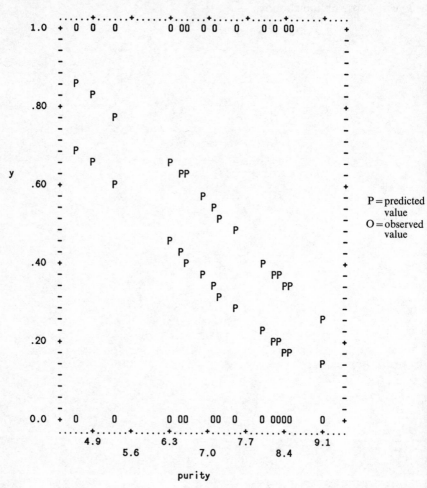

P = predicted value
O = observed value

Output H.3

Maximized log likelihood, estimates of parameters and s.e.s for full model.
BMDPLR

```
TOTAL NUMBER OF RESPONSES USED IN THE ANALYSIS      44
                        f        . . . . . .        18
                        nf       . . . . . .        26

                   LOG LIKELIHOOD =    -26.406
GOODNESS OF FIT CHI-SQ   (2*0*LN(0/E)) =   38.582   D.F.=  29   P-VALUE= 0.110
GOODNESS OF FIT CHI-SQ   ( D. HOSMER ) =    4.196   D.F.=   8   P-VALUE= 0.839
GOODNESS OF FIT CHI-SQ   ( C.C.BROWN ) =    2.715   D.F.=   2   P-VALUE= 0.257

                                      STANDARD
        TERM        COEFFICIENT        ERROR     COEFF/S.E.

purity               -0.60421         0.2838      -2.129
process               0.43225         0.3358       1.287
CONSTANT              4.0286          2.092        1.926
```

Example I
Growth of bones from chick embryos

Description of data

Table I.1 gives data on the growth of bones from seven-day-old chick embryos after cultivation over a nutrient chemical medium (Biggers and Heyner, 1961). The observations are of log dry weight (μg). Two bones were available from each embryo and the experiment was therefore set out in a (balanced) incomplete block design with two units per block. C denotes the complete medium with about 30 ingredients in carefully controlled quantities. The five other media were obtained by omitting a single amino acid, e.g., His$^-$ is a medium without L-histidine, etc. The treatment pairs were randomized, but the following results are given in systematic order. Interest lies in comparing the effects of omitting the various amino acids.

Table I.1. Log_{10}(dry weight) of tibiotarsi from seven-day-old chick embryos

Embryo 1	C	2.51: His$^-$	2.15	9 His$^-$	2.32: Lys$^-$	2.53
2	C	2.49: Arg$^-$	2.23	10 Arg$^-$	2.15: Thr$^-$	2.23
3	C	2.54: Thr$^-$	2.26	11 Arg$^-$	2.34: Val$^-$	2.15
4	C	2.58: Val$^-$	2.15	12 Arg$^-$	2.30: Lys$^-$	2.49
5	C	2.65: Lys$^-$	2.41	13 Thr$^-$	2.20: Val$^-$	2.18
6	His$^-$	2.11: Arg$^-$	1.90	14 Thr$^-$	2.26: Lys$^-$	2.43
7	His$^-$	2.28: Thr$^-$	2.11	15 Val$^-$	2.28: Lys$^-$	2.56
8	His$^-$	2.15: Val$^-$	1.70			

The analysis

For the within-block analysis, differences $W_i = Y_{i1} - Y_{i2}$ ($i=1, \ldots, 15$) are calculated within the program and fitted to the model

$$E(\mathbf{W}) = \mathbf{X}\boldsymbol{\theta} \qquad \text{(I.4, App. Stat.)}$$

where $\boldsymbol{\theta}^{\mathrm{T}} = (\theta_1, \ldots, \theta_5)$ and $\theta_k = \tau_c - \tau_k$ is the difference in expected response for treatment k compared with the complete medium. If the data are input in the form suggested, the elements of $\mathbf{X}$ are equal simply to minus the values in

56

columns 2 to 6; a regression model with zero constant is fitted. Residuals are printed and plotted.

For the between-block analysis, the block totals $Z_i = Y_{i1} + Y_{i2}$ are computed. Instead of the model

$$E(Z_i) = 2\mu + \tau_{k_{1i}} + \tau_{k_{2i}} \qquad \text{(I.5, App. Stat.)}$$

we can write

$$E(Z_i) = \gamma_{k_{1i}} + \gamma_{k_{2i}}$$

incorporating the constant μ into the γ_k's. Then if $E(\mathbf{Z}) = \mathbf{U}\gamma$ the elements of $\mathbf{U}$ are equal to 1 wherever ± 1 occurs in columns 1 to 6 of the input data and a regression with zero constant is required. The estimates $\tilde{\theta}_k$ of Table I.2, *App. Stat.* are equal to $\tilde{\gamma}_c - \tilde{\gamma}_k$; the correlation matrix of the regression coefficients is required in order to obtain s.e.$(\tilde{\theta}_k)$.

Program

BMDP1R is used for the within-block and for the between-block analyses. For the within-block analysis the instructions are:

```
/problem     title = 'example i. chick embryos.
             within-blocks. bmdp1r'.
/input       variables = 8.  format = free.
/variable    names = c, h, a, t, v, l, y1, y2.
             add = new.
/transform   w = y2 - y1.                          within-block differences
/regress     dependent = w.  independent = 2 to 6.
             type = zero.                          constant set to zero
/print       data.                                print data, predicted values and
/plot        normal.  residuals.                      residuals
/end                                              normal plot of residuals
                                                  residual v. predicted

1 -1 0 0 0 0 2.51 2.15                            15 lines of data
1 0 -1 0 0 0 2.49 2.23                            In cols 1–6 code 1st and 2nd treat-
                                                  ments as 1 and −1, resp., e.g.
         . . .                                    cols 1–6 for line 15 read
                                                  0 0 0 0 1 −1
```

Output I.1 shows the estimates $\tilde{\theta}_k$ and s.e.s (as in Table I.2, *App. Stat.*). Output I.2 shows the residuals plotted against the predicted values, and Output I.3 the normal probability plot. Included in the output, but not shown here, is a list of the predicted values and residuals.

For the between-block analysis the instructions are:

```
/problem     title = 'example i. chick embryos.
             between-blocks. bmdp1r'.
/input       variables = 8. format = free.
/variable    names = c, h, a, t, v, l, y1, y2.
             add = new.
/transform   z = y1 + y2.
             for  k = 2 to 6.
             % x(k) = x(k)**2. %                    square data cols 2–6
```

```
/regress      dependent = z.  independent = 1 to 6.
              type = zero.
/print        data.  rreg.                          corr. matrix of estimates
/plot         normal.  residuals.
/end
1 -1 0 0 0 0 2.51 2.15                              15 lines of data as before
1 0 -1 0 0 0 2.49 2.23

      ...
```

Output I.4 shows the estimates $\hat{\gamma}_k$. The estimates $\tilde{\theta}_k$ of Table I.2, *App. Stat.* are obtained as $\tilde{\theta}_1 = \hat{\gamma}_C - \hat{\gamma}_H = 2.56050 - 2.00800 = 0.55250$, $\tilde{\theta}_2 = \hat{\gamma}_C - \hat{\gamma}_A = 2.56050 - 2.16550 = 0.39500$, etc.

The correlation matrix of the parameter estimates (not shown here) has all off-diagonal elements equal to -0.1111 and hence the estimated s.e. of each $\tilde{\theta}_k$ is equal to s.e.$(\hat{\gamma}_k)\sqrt{2.2222} = 0.1250$, with 9 d.f. (Note the d.f. are incorrectly taken as 8 in early printings of *App. Stat.*)

The within-block estimates $\tilde{\theta}_k$ and the between-block estimates $\bar{\theta}_k$ can be pooled by hand, weighting each inversely proportionally to its estimated variance, to obtain the pooled estimates θ_k^* given in Table I.2, *App. Stat.*

Other program

BMDP2R may be used, with the **set** option being used to permit non-stepwise entry of variables.

Suggested further work

(i) How would you examine for homogeneity of variance?
(ii) Reproduce the within-block analysis by fitting a model to the original data (not the differences) in which each observation has an expected value equal to a constant depending on the embryo plus a constant depending on the treatment. Note that care is needed to avoid a singular parameterization.

Output I.1

Within-block analysis. Estimates of parameters θ_k and s.e.s. BMDP1R

```
ANALYSIS OF VARIANCE
               SUM OF SQUARES     DF     MEAN SQUARE      F RATIO    P(TAIL)
  REGRESSION          0.8924      5          0.1785       13.562     0.0003
  RESIDUAL            0.1316     10          0.0132
```

VARIABLE		COEFFICIENT	STD. ERROR	STD. REG COEFF	T	P(2 TAIL)	TOLERANCE
h	2	0.2183	0.0662	0.48	3.30	0.01	0.6000
a	3	0.3533	0.0662	0.78	5.33	0.00	0.6000
t	4	0.3483	0.0662	0.77	5.26	0.00	0.6000
v	5	0.4900	0.0662	1.08	7.40	0.00	0.6000
l	6	0.1600	0.0662	0.35	2.42	0.04	0.6000

Output I.2

Within-block analysis. Residuals versus predicted values. BMDP1R

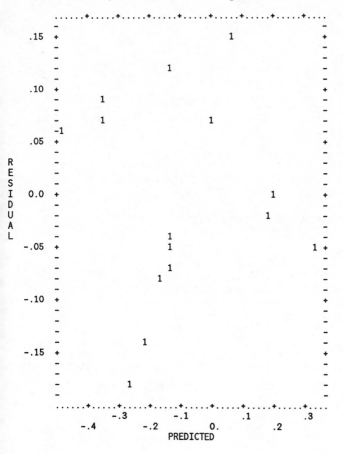

Output I.3

Within-block analysis. Normal probability plot. BMDP1R

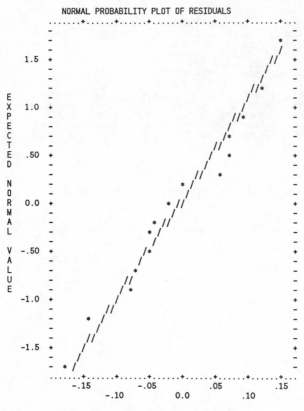

VALUES FROM NORMAL DISTRIBUTION WOULD LIE
ON THE LINE INDICATED BY THE SYMBOL / .

Output I.4

Between-block analysis. Estimates of parameters γ_k and s.e.s. BMDP1R

ANALYSIS OF VARIANCE

	SUM OF SQUARES	DF	MEAN SQUARE	F RATIO	P(TAIL)
REGRESSION	315.3244	6	52.5541	1701.041	0.0000
RESIDUAL	0.2781	9	0.0309		

VARIABLE		COEFFICIENT	STD. ERROR	STD. REG COEFF	T	P(2 TAIL)	TOLERANCE
c	1	2.5605	0.0834	0.32	30.71	0.00	0.8889
h	2	2.0080	0.0834	0.25	24.08	0.00	0.8889
a	3	2.1655	0.0834	0.27	25.97	0.00	0.8889
t	4	2.2280	0.0834	0.28	26.72	0.00	0.8889
v	5	2.1405	0.0834	0.27	25.67	0.00	0.8889
l	6	2.6255	0.0834	0.33	31.49	0.00	0.8889

Example J Factorial experiment on cycles to failure of worsted yarn

Description of data

In an unpublished report to the Technical Committee, International Wool Textile Organization, A. Barella and A. Sust gave the data in the first four columns of Table J.1, concerning the number of cycles to failure of lengths of worsted yarn under cycles of repeated loading. The three factors which varied over levels specified in coded form in the first three columns, are:

x_1, length of test specimen   (250, 300, 350 mm);
x_2, amplitude of loading cycle   (8, 9, 10 mm);
x_3, load   (40, 45, 50 g).

The analysis

The problem would be computationally straightforward if analysed as a 3^3 experiment with equal spacing of the levels for each factor, but taking log Y as dependent on log x_1, log x_2 and log x_3, as in *App. Stat.*, destroys the advantages of the equal spacing on x_1, x_2 and x_3.

The analysis is most simply tackled by multiple regression, fitting a model with linear, linear × linear and quadratic terms. Of course, if only linear terms are to be included many of the special points dealt with below will not arise. If using BMDP1R to fit a model with terms log x_i, (log x_i)^2, $i = 1, 2, 3$, it is helpful to centralize log x_i about its mean, as in Example G, as the program will automatically exclude variables having a high correlation with others in the model.

For an analysis corresponding to Table J.3, *App. Stat.* in which all terms are included and for which all terms can be interpreted separately, we need to orthogonalize the linear and quadratic polynomials. For three unequally spaced points z_i $(i = 1, 2, 3)$ such that $\Sigma z_i = 0$, the values of the linear and quadratic orthogonal polynomials are given, respectively, by z_i and $z_i^2 + b z_i +$

61

Table J.1. Cycles to failure, transformed values, fitted
values and residuals

			Cycles	Log cycles		
x_1	x_2	x_3	obs	obs	fitted	resid.
−1	−1	−1	674	6.51	6.52	−0.01
−1	−1	0	370	5.91	6.11	−0.20
−1	−1	1	292	5.68	5.74	−0.06
−1	0	−1	338	5.82	5.85	−0.03
−1	0	0	266	5.58	5.44	0.14
−1	0	1	210	5.35	5.07	0.28
−1	1	−1	170	5.14	5.26	−0.12
−1	1	0	118	4.77	4.84	−0.07
−1	1	1	90	4.50	4.48	0.02
0	−1	−1	1414	7.25	7.42	−0.17
0	−1	0	1198	7.09	7.01	0.08
0	−1	1	634	6.45	6.64	−0.19
0	0	−1	1022	6.93	6.76	0.17
0	0	0	620	6.43	6.34	0.09
0	0	1	438	6.08	5.97	0.11
0	1	−1	442	6.09	6.16	−0.07
0	1	0	332	5.81	5.75	0.06
0	1	1	220	5.39	5.38	0.01
1	−1	−1	3636	8.20	8.18	0.02
1	−1	0	3184	8.07	7.77	0.30
1	−1	1	2000	7.60	7.40	0.20
1	0	−1	1568	7.36	7.52	−0.16
1	0	0	1070	6.98	7.11	−0.13
1	0	1	566	6.34	6.74	−0.40
1	1	−1	1140	7.04	6.92	0.12
1	1	0	884	6.78	6.51	0.27
1	1	1	360	5.89	6.14	−0.25

x_1, length; x_2, amplitude of loading cycle; x_3, load.

c, where $b = -\Sigma z_i^3/\Sigma z_i^2$ and $c = -\frac{1}{3}\Sigma z_i^2$. These may be calculated in the
transform paragraph although since few values for z_i, b and c are required we
choose here to calculate these quantities by hand and write them into the
transform paragraph; note that z_i, b and c, since working in logarithms, are
identical for amplitude and load. The required values are:

	z_1	z_2	z_3	b	c
Length	-0.1729	0.0094	0.1635	$0.014\,056$	$-0.018\,905$
Amplitude ⎫ Load ⎭	-0.1136	0.0041	0.1095	$0.006\,142$	$-0.008\,304$

These are written into the program instructions as shown below.

Program

BMDP9D is used to compute two-way means (Table J.2, *App. Stat.*) for initial inspection. The instructions are:

```
/problem     title = 'example j.  3x3x3 expt.
             two-way means.  bmdp9d'.
/input       variables = 4.  format = free.
/variable    names = length, amplitude, weight, y.    names with different initial letters
/category    codes(length to weight) = -1, 0, 1.
             names(length to weight) =                 names for labelling output
             '-1','0','+1'.
/transform   y = ln(y).                                log cycles
/tabulate    grouping = length, amplitude, weight.
             margin = 'la.', 'l.w', '.aw'.             two-way tables
/end
-1 -1 -1 674                                           27 lines of data
-1 -1 0 370

. . .
```

Names assigned in the **variable** paragraph should begin with different letters in order that the two-way tables may be requested as shown in the **margin** statement of, the **tabulate** paragraph; hence instead of the name 'load' for x_3 we use 'weight'. Output J.1 shows the two-way means for amplitude × weight; the tables for length × amplitude and length × weight are not shown here.

BMDP1R is used for a multiple regression of log cycles upon log x_1, log x_2 and log x_3, fitting linear, quadratic and linear × linear terms. The linear and quadratic orthogonal polynomial values are computed from the values of z_1, z_2, z_3, b and c given above using the **recode** instruction in the **transform** paragraph. The instructions for BMDP1R are:

```
/problem     title = 'example j.  3x3x3 expt.
             bmdp1r.'.
/input       variables = 4.  format = free.
/variable    names = x1, x2, x3, y.  add = new.
/transform   y = ln(y).                                log cycles
             b1 = 0.014056.  b2 = 0.006142.            coefficients for quadratic poly-
             c1 = -0.018905.  c2 = -0.008304.          nomial
             x1 = rec(x1, -1, -0.1729, 0, 0.0094,      values of z1, z2, z3
             1, 0.1635).
             x2 = rec(x2, -1, -0.1136, 0, 0.0041,
             1, 0.1095).
             x3 = rec(x3, -1, -0.1136, 0, 0.0041,
             1, 0.1095).
```

```
                x11 = x1**2 + b1*x1 +c1.        compute  linear  and  quadratic
                x22 = x2**2 + b2*x2 + c2.          values
                x33 = x3**2 + b2*x3 + c2.
                x12 = x1*x2.                    compute linear × linear values
                x13 = x1*x3.
                x23 = x2*x3.
/regress        dependent = y.
                independent = x1, x2, x3, x12, x13,
                x23, x11, x22, x33.
/print          data.                          print data, predicted, residuals
/plot           normal.                        . normal probability plot
/end
-1 -1 -1 674                                    27 lines of data as before
-1 -1 0 370

    . . .
```

The linear terms are denoted by **x1, x2, x3**, the linear × linear terms by **x12**, **x13, x23** and the quadratic terms by **x11, x22** and **x33**. Output J.2 shows estimated coefficients and s.e.s of the fitted model; this is equivalent to the analysis of variance given in Table J.3, *App. Stat.* Only the linear terms are significant. Output J.3 shows a normal probability plot of the residuals. The output includes also a list of the residuals (not shown here).

Other program

BMDP2R may be used, with the **set** option being used to permit non-stepwise entry of variables.

Suggested further work

Compare the fit of log Y on x_1, x_2. x_3 with that of log Y on log x_1, log x_2, log x_3 (using linear terms only).

Output J.1

Two-way means for amplitude × weight. BMDP9D

```
          ALL
CELL      length    amplitud   weight
NUMBER    1         2          3
   1               -1         -1
   2                0         -1
   3               +1         -1
   4               -1          0
   5                0          0
   6               +1          0
   7               -1         +1
   8                0         +1
   9               +1         +1
```

DESCRIPTIVE STATISTICS FOR NON-EMPTY CELLS ON MARGINALS '.aw'

```
 CELL      VARIABLE   4 y
NUMBER     FREQ.    MEAN      STD.DEV.

   1        3.      7.32202   0.8448
   2        3.      6.70337   0.7919
   3        3.      6.08863   0.9515
   -------------------------------
   4        3.      7.02260   1.0777
   5        3.      6.32954   0.7013
   6        3.      5.78676   1.0070
   -------------------------------
   7        3.      6.57657   0.9681
   8        3.      5.92264   0.5146
   9        3.      5.25985   0.7028
```

Similar tables are output for length × amplitude and length × weight.

Output J.2

Estimates and s.e.s. BMDP1R

ANALYSIS OF VARIANCE

	SUM OF SQUARES	DF	MEAN SQUARE	F RATIO	P(TAIL)
REGRESSION	22.5630	9	2.5070	66.520	0.0000
RESIDUAL	0.6407	17	0.0377		

VARIABLE		COEFFICIENT	STD. ERROR	STD. REG COEFF	T	P(2 TAIL)	TOLERANCE
INTERCEPT		6.33466					
x1	1	4.95142	0.27173	0.734	18.222	0.0000	1.00000
x2	2	-5.65484	0.40999	-0.556	-13.793	0.0000	1.00000
x3	3	-3.50388	0.40999	-0.344	-8.546	0.0000	1.00000
x11	9	-0.56669	2.82451	-0.008	-0.201	0.8434	1.00000
x22	10	-0.85200	6.39187	-0.005	-0.133	0.8955	1.00000
x33	11	-7.18180	6.39187	-0.045	-1.124	0.2768	1.00000
x12	12	-2.18317	2.98186	-0.030	-0.732	0.4740	1.00000
x13	13	-3.49334	2.98186	-0.047	-1.172	0.2575	1.00000
x23	14	-1.63938	4.49916	-0.015	-0.364	0.7201	1.00000

Output J.3

Normal probability plot. BMDP1R

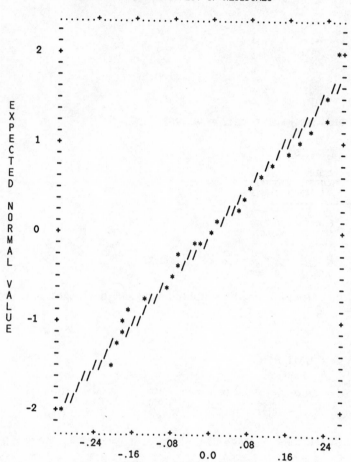

NORMAL PROBABILITY PLOT OF RESIDUALS

VALUES FROM NORMAL DISTRIBUTION WOULD LIE
ON THE LINE INDICATED BY THE SYMBOL / .

Example K   Factorial experiment on diets for chickens

Description of data

An experiment comparing 12 methods of feeding chickens (Duckworth and Carpenter; see John and Quenouille, 1977) was done independently in two replicates arranged in different houses. The treatments, forming a $3 \times 2 \times 2$ factorial, were 'form of protein', 'level of protein', 'level of fish solubles'. The data are given in Table K.1.

Table K.1. Total weights of 16 six-week-old chicks (g)

Protein	Level of protein	Level of fish solubles	House I	II
Groundnut	0	0	6559	6292
		1	7075	6779
	1	0	6564	6622
		1	7528	6856
	2	0	6738	6444
		1	7333	6361
Soyabean	0	0	7094	7053
		1	8005	7657
	1	0	6943	6249
		1	7359	7292
	2	0	6748	6422
		1	6764	6560

The analysis

The design is a balanced factorial experiment with 2 proteins × 3 levels of protein × 2 fish solubles, replicated in 2 houses. The whole analysis can be done by BMDP4V with a single submission of the data. The program below has three **analysis** paragraphs; these (or others) can be added sequentially if the program is being run interactively or, as shown below, can all be submitted together. The **print** paragraph gives all one- and two-way means (Table K.2,

67

App. Stat.). The **analysis** paragraphs compute analysis of variance tables for

(i) the full $2 \times 3 \times 2 \times 2$ factorial design;
(ii) $2 \times 3 \times 2$ factorial within each house; and
(iii) reduced model in which the interactions with houses are pooled.

The output from (i) and (ii) gives the analysis of variance of Table K.3(a), *App. Stat.* and that from (iii) gives the analysis of variance of Table K.3(b), *App. Stat.*

Program

BMDP4V is used. The instructions are:

```
/problem     title = 'example k.
             diets for chickens.  bmdp4v'.
/input       variables = 5.  format = free.
/variable    names  =  protein, levelp, fish,
             house, y.
/group       codes(1) = 1, 2.
             names(1) = grnut, soya.
             codes(2) = 0, 1, 2.
             names(2) = lp0, lp1, lp2.        labels for output
             codes(3) = 0, 1.
             names(3) = fish0, fish1.
             codes(4) = 1, 2.
             names(4) = house1, house2.
/between     factors = protein, levelp, fish,
             house.
/weight      between = equal.
/print       margins = 1, 2.                  1- and 2-way means
/end
1 0 0 1 6559                                   24 lines of data
1 0 0 2 6292                                   Code: groundnut = 1,
                                                     soyabean = 2 (col. 1)
   . . .

/end
analysis     procedure = factorial.  unisum. /    2 × 3 × 2 × 2 factorial
analysis     procedure = simple.  hold = house.    2 × 3 × 2 factorial within each
             unisum. /                             house
analysis     procedure = structure.
             bform = 'house + protein*levelp*fish'.  reduced model
             unisum. /
end/
```

Note: BMDP4V requires the data to be inserted before the **analysis** statements.

Output K.1 shows the one- and two-way means, with s.e.s; Outputs K.2, K.3 and K.4 give the analysis of variance for the full $2 \times 3 \times 2 \times 2$ factorial, the $2 \times 3 \times 2$ factorial within house I and the reduced model in which interactions with houses are pooled. The protein × level of protein interaction is evident in Output K.4.

Other programs

BMDP9D gives one- and two-way marginal means.

BMDP2V computes analysis of variance for full factorial, or for specified model; it will also print fitted values and residuals.

BMDP8V computes one- and two-way marginal means and analysis of variance for factorial designs; it needs careful organization of the data (see Examples Q, R).

BMDP1R is a multiple regression program; if protein and fish-soluble levels and each coded as $-1, 1$ and the three levels of protein as $-1, 0, 1$, then all the factorial effects can be estimated by single degree of freedom orthogonal contrasts. BMDP2R may likewise be used if the **set** option is used to permit non-stepwise entry of variables.

Suggested further work

(i) Repeat the analysis using log weight. Comment.

(ii) Carry out the analysis using BMDP1R as suggested above.

Output K.1

One- and two-way marginal means. BMDP4V

LEVEL 1 MARGINALS

FACTOR	LEVEL	COUNT	MEAN	STDERROR	STD_DEV	WTD_MEAN	MAXIMUM	MINIMUM
protein	grnut	12	6762.5833	110.2099	381.7782	6762.5833	7528.0000	6292.0000
	soya	12	7012.1667	147.7052	511.6659	7012.1667	8005.0000	6249.0000
levelp	lp0	8	7064.2500	196.8731	556.8413	7064.2500	8005.0000	6292.0000
	lp1	8	6926.6250	156.4577	442.5291	6926.6250	7528.0000	6249.0000
	lp2	8	6671.2500	110.1348	311.5082	6671.2500	7333.0000	6361.0000
fish	fish0	12	6644.0000	80.8176	279.9604	6644.0000	7094.0000	6249.0000
	fish1	12	7130.7500	139.6197	483.6569	7130.7500	8005.0000	6361.0000
house	house1	12	7059.1667	125.4296	434.5009	7059.1667	8005.0000	6559.0000
	house2	12	6715.5833	125.2838	433.9957	6715.5833	7657.0000	6249.0000

LEVEL 2 MARGINALS

===

FACTOR	LEVEL
protein ==>	grnut

FACTOR	LEVEL	COUNT	MEAN	STDERROR	STD_DEV	WTD_MEAN	MAXIMUM	MINIMUM
levelp	lp0	4	6676.2500	166.0710	332.1419	6676.2500	7075.0000	6292.0000
	lp1	4	6892.5000	221.0345	442.0690	6892.5000	7528.0000	6564.0000
	lp2	4	6719.0000	220.0655	440.1310	6719.0000	7333.0000	6361.0000
fish	fish0	6	6536.5000	62.5863	153.3046	6536.5000	6738.0000	6292.0000
	fish1	6	6988.6667	170.5311	417.7141	6988.6667	7528.0000	6361.0000
house	house1	6	6966.1667	167.4963	410.2806	6966.1667	7528.0000	6559.0000
	house2	6	6559.0000	93.8609	229.9113	6559.0000	6856.0000	6292.0000

===

FACTOR	LEVEL
protein ==>	soya

FACTOR	LEVEL	COUNT	MEAN	STDERROR	STD_DEV	WTD_MEAN	MAXIMUM	MINIMUM
levelp	lp0	4	7452.2500	230.0722	460.1445	7452.2500	8005.0000	7053.0000
	lp1	4	6960.7500	254.1707	508.3413	6960.7500	7359.0000	6249.0000
	lp2	4	6623.5000	81.5858	163.1717	6623.5000	6764.0000	6422.0000
fish	fish0	6	6751.5000	142.1231	348.1291	6751.5000	7094.0000	6249.0000
	fish1	6	7272.8333	220.4909	540.0901	7272.8333	8005.0000	6560.0000
house	house1	6	7152.1667	194.1874	475.6601	7152.1667	8005.0000	6748.0000
	house2	6	6872.1667	224.6003	550.1561	6872.1667	7657.0000	6249.0000

===

Similar tables are output for other two-way means.

Output K.2

Analysis of variance for $2 \times 3 \times 2 \times 2$ factorial. BMDP4V

SOURCE	SUM OF SQUARES	DF	MEAN SQUARE
OVALL: GRAND MEAN	1.138462425E+9	1	1.1384624E+9
p: protein	373751.04167	1	373751.04167
l: levelp	636282.75000	2	318141.37500
f: fish	1421553.37500	1	1421553.37500
h: house	708297.04167	1	708297.04167
pl	858157.58333	2	429078.79167
pf	7176.04167	1	7176.04167
ph	24257.04167	1	24257.04167
lf	308887.75000	2	154443.87500
lh	44521.08333	2	22260.54167
fh	41251.04167	1	41251.04167
plf	50127.58333	2	25063.79167
plh	49940.58333	2	24970.29167
pfh	147110.04167	1	147110.04167
lfh	12829.08333	2	6414.54167
plfh	172731.58333	2	86365.79167
ERROR	0.00000	0	0.00000

Output K.3

Analysis of variance for $2 \times 3 \times 2$ factorial within House I. BMDP4V

SOURCE	SUM OF SQUARES	DF	MEAN SQUARE
OVALL.A: MEAN AT hou	5.979820083E+8	1	5.9798201E+8
p.A: protein AT hous	103788.00000	1	103788.00000
l.A: levelp AT house	174595.16667	2	87297.58333
f.A: fish AT housel	973560.33333	1	973560.33333
pl.A	521913.50000	2	260956.75000
pf.A	44652.00000	1	44652.00000
lf.A	104952.16667	2	52476.08333
plf.A	153240.50000	2	76620.25000
ERROR	0.00000	0	0.00000

A similar table is output for House II.

Output K.4

Analysis of variance for reduced model, interactions with houses pooled. BMDP4V

SOURCE	SUM OF SQUARES	DF	MEAN SQUARE
OVALL: GRAND MEAN	1.138462425E+9	1	1.1384624E+9
h:house	708297.04167	1	708297.04167
p:protein	373751.04167	1	373751.04167
l:levelp	636282.75000	2	318141.37500
f:fish	1421553.37500	1	1421553.37500
pl	858157.58333	2	429078.79167
pf	7176.04167	1	7176.04167
lf	308887.75000	2	154443.87500
plf	50127.58333	2	25063.79167
ERROR	0.00000	0	0.00000

Example L   Binary preference data for detergent use

Description of data

Table L.1 (Ries and Smith, 1963) compares two detergents, a new product X and a standard product M. Each individual expresses a preference between X and M. In the table, Y_j is the number of individuals out of n_j in 'cell' j who prefer X, the remaining $n_j - Y_j$ preferring M. The individuals are classified by three factors, water softness at three levels, temperature at two levels, and a factor whose two levels correspond to previous experience and no previous experience with M. The object is to study how preferences for X vary.

Table L.1. Number Y_j of preferences for brand X out of n_j individuals

Water softness		M previous non-user		M previous user	
		Temperature		Temperature	
		Low	High	Low	High
Hard	Y_j	68	42	37	24
	n_j	110	72	89	67
Medium	Y_j	66	33	47	23
	n_j	116	56	102	70
Soft	Y_j	63	29	57	19
	n_j	116	56	106	48

The analysis

The discussion in *App. Stat.* points out the range in proportions of preferences for brand X (Y_j/n_j) is such that an analysis of the untransformed proportions is reasonable and has the advantage of direct interpretation. If, however, the results are to be compared with another data set having proportions at a higher or lower level, then a logistic analysis is likely to be preferable. We give programs for both methods of analysis.

BMDP4V is useful for analysis of the proportions in the $3 \times 2 \times 2$ factorial arrangement; it prints one- and two-way marginal means in addition to computing the analysis of variance table. BMDPLR fits a logistic model.

Programs

BMDP4V computes an analysis of variance of the proportions. The proportions are calculated in the **transform** paragraph. The three levels of water softness are treated as equally spaced with linear and quadratic components extracted; orthogonal coefficients $-1, 0, 1$ and $1, -2, 1$ are written into the **design** paragraph. The instructions are:

```
/problem      title = 'example I. detergent
              preference. bmdp4v.'.
/input        variables = 5.  format = free.
/variable     names  =  wsoft, muser, temp, y, n.
              add = new.
              use = wsoft, muser, temp, prop.
/group        codes(wsoft) = 1, 2, 3.
              names(wsoft) = hard, medium, soft.
              codes(muser) = 1, 2.
              names(muser) = m_nonuse, m_use.
              codes(temp) = 1, 2.
              names(temp) = low, high.
/transform    prop = y/n.                        compute proportions
/between      factors = wsoft, muser, temp.
/weights      between = equal.
/print        margins = all.
/end
1 1 1 68 110                                     12 lines of data
1 1 2 42 72                                      Code: wsoftness (hard,med,soft)
    ...                                                    =1,2,3
                                                       M(nonuser,user)=1,2
                                                       Temp(low,high)=1,2
/end
design        factor = wsoft.
              type = between, regression.
              code = read.  values = -1, 0, 1.   linear coefficients
              name = linws. /
design        factor = wsoft.  values = 1, -2, 1.  quadratic coefficients
              name = quadws. /
analysis      procedure = structure.  unisum.
              bform = '(linws + quadws)*         factorial, with lin. and quad. for w-
              muser*temp'. /                     softness
end/
```

Note: BMDP4V requires the data to be inserted before the **design** and **analysis** statements.
Output L.1 shows the one- and two-way means (as in Table L.3, *App. Stat.*). Output L.2 gives the analysis of variance corresponding to Table L.4(a), *App. Stat.* The theoretical error variance, with which the given mean squares should be compared, is simply calculated by hand as the average of $\{Y_j(n_j - Y_j)\}/\{n_j^2(n_j - 1)\}$.

Given there is some evidence of interaction with previous usage of M, we need to repeat the analysis separately for users and non-users. For non-users (or users) the above instructions should be amended as follows. To the **transform** paragraph add the statement

use = muser eq 1. (or **muser eq 2.** for users).

Amend the **between** paragraph statement to

factors = wsoft, temp.

and the **bform** statement in the **analysis** paragraph to

bform = '(linws + quadws)∗temp'./

Two separate runs will give the required analysis. If wished they may instead be done in one submission using the **for–%** option; see note 9(ii) in the introductory section (p. 9).

For a logistic regression using BMDPLR the instructions are:

```
/problem     title = 'example I. detergent preference.
             bmdplr'.
/input       variables = 5.  format = free.
/variable    names =    wsoft, muser, temp, y, n.
             add = new.
/group       code(wsoft) = 1, 2, 3.
             names(wsoft) = hard, medium, soft.
             code(muser) = 1, 2.
             names(muser) = m_nonuse, m_use.
             code(temp) = 1, 2̄.
             names(temp) = low, high.
/transform   ls = wsoft - 2.                      ls = linear coeff. wsoft.
             qs = 3∗ls∗∗2 - 2.                    qs = quadratic coeff.
/regress     count = n.
             scount = y.
             model = temp∗muser∗ls, temp∗muser∗qs.
             interval = ls, qs.                   ls,qs continuous variables
/end
1 1 1 68 110                                      12 lines of data as before
1 1 2 42 72

    . . .
```

The **model** statement **temp∗muser∗ls** in the **regress** paragraph specifies the main effects, two-factor interactions and three-factor interaction involving the linear component of water softness; **temp∗muser∗qs** involves the quadratic component. The two statements together give the usual breakdown for the full $3 \times 2 \times 2$ factorial. Output L.3 shows the log likelihood and estimated coefficients with s.e.s.

To fit models separately to non-users and users of M, we modify the **model** statement, deleting **muser**, and add a **use** statement to the **transform** paragraph. Instructions for non-users are given below. For users of M change the **use** sentence to read use = **muser eq 2.**

```
/problem     title = 'example I.
             detergent preference.
             non-users of m.  bmdplr'.
/input       variables = 5.  format = free.
/variable    names = wsoft, muser, temp, y, n.
             add = new.
```

```
/group        code(wsoft) = 1, 2, 3.
              names(wsoft) = hard, medium, soft.
              code(muser) = 1, 2.
              names(muser) = m_nonuse, m_use.
              code(temp) = 1, 2.
              names(temp) = low, high.
/transform    use = muser eq 1.
              ls = wsoft - 2.
              qs = 3*ls**2 - 2.
/regress      count = n.
              scount = y.
              model = temp*ls, temp*qs.
              interval = ls, qs.
/end
1 1 1 68 110                          12 lines of data as before
1 1 2 42 72
   . . .
```

Note: the **for–%** option may be used to make the program repeat automatically for non-users and users of M; see note 9(ii) in the introductory section (p. 9).

Output L.4 gives the results for previous non-users of M. The logistic regression and the analysis of proportions lead to similar conclusions.

Other programs

BMDP2V will compute an analysis of variance for the $3 \times 2 \times 2$ factorial and print out residuals. BMDP1R is a regression program and could be used to estimate all the single degree of freedom contrasts instead of computing an analysis of variance. It will also permit examination of residuals. BMDP2R can be used if the **set** option is used to permit non-stepwise entry of variables.

Suggested further work

(i) Use BMDP1R (or BMDP2R with **set** option) to analyse the proportions. Examine also the residuals.

(ii) How in general would one decide whether linear, or logistic, or some other model is preferable?

Output L.1

One- and two-way means of proportions. BMDP4V

LEVEL 1 MARGINALS

FACTOR	LEVEL	COUNT	MEAN	STDERROR	STD_DEV	WTD_MEAN	MAXIMUM	MINIMUM
wsoft	hard	4	0.4939	0.0632	0.1264	0.4939	0.6182	0.3582
	medium	4	0.4869	0.0598	0.1197	0.4869	0.5893	0.3286
	soft	4	0.4986	0.0347	0.0694	0.4986	0.5431	0.3958
muser	m_nonuse	6	0.5701	0.0145	0.0355	0.5701	0.6182	0.5179
	m_use	6	0.4161	0.0307	0.0751	0.4161	0.5377	0.3286
temp	low	6	0.5241	0.0301	0.0737	0.5241	0.6182	0.4157
	high	6	0.4622	0.0473	0.1158	0.4622	0.5893	0.3286

LEVEL 2 MARGINALS

===
FACTOR	LEVEL
wsoft	hard
==>	

FACTOR	LEVEL	COUNT	MEAN	STDERROR	STD_DEV	WTD_MEAN	MAXIMUM	MINIMUM
muser	m_nonuse	2	0.6008	0.0174	0.0246	0.6008	0.6182	0.5833
	m_use	2	0.3870	0.0288	0.0407	0.3870	0.4157	0.3582
temp	low	2	0.5170	0.1012	0.1432	0.5170	0.6182	0.4157
	high	2	0.4708	0.1126	0.1592	0.4708	0.5833	0.3582

===
FACTOR	LEVEL
wsoft	medium
==>	

FACTOR	LEVEL	COUNT	MEAN	STDERROR	STD_DEV	WTD_MEAN	MAXIMUM	MINIMUM
muser	m_nonuse	2	0.5791	0.0102	0.0144	0.5791	0.5893	0.5690
	m_use	2	0.3947	0.0661	0.0935	0.3947	0.4608	0.3286
temp	low	2	0.5149	0.0541	0.0765	0.5149	0.5690	0.4608
	high	2	0.4589	0.1304	0.1844	0.4589	0.5893	0.3286

===

Similar tables are output for other two-way means.

Output L.2

Model specification and analysis of variance for $3 \times 2 \times 2$ factorial. BMDP4V

INDEX OF NAMES IN FORMULA

```
l:linws
q:quadws
m:muser
t:temp
```

DECODED FORMULA

```
OVALL + l + q + m + t + lm + qm + lt + qt + mt + lmt + qmt
```

Output L.2 (continued)

SOURCE	SUM OF SQUARES	DF	MEAN SQUARE
linws	0.00005	1	0.00005
quadws	0.00023	1	0.00023
OVALL: GRAND MEAN	2.91816	1	2.91816
m:muser	0.07113	1	0.07113
t:temp	0.01150	1	0.01150
lm	0.01126	1	0.01126
qm	0.00139	1	0.00139
lt	0.00070	1	0.00070
qt	0.00005	1	0.00005
mt	0.00710	1	0.00710
lmt	0.00110	1	0.00110
qmt	0.00114	1	0.00114
ERROR	0.00000	0	0.00000

Output L.3

Logistic regression. Full data. BMDPLR

STEP NUMBER   0

LOG LIKELIHOOD =   -682.248

TERM	COEFFICIENT	STANDARD ERROR	COEFF/S.E.
temp	-0.12815	0.6725E-01	-1.905
muser	-0.31402	0.6725E-01	-4.669
t*m	-0.10092	0.6725E-01	-1.501
ls	0.97297E-02	0.8274E-01	0.1176
t*l	-0.35911E-01	0.8274E-01	-0.4340
m*l	0.15323	0.8274E-01	1.852
t*m*l	-0.46906E-01	0.8274E-01	-0.5669
qs	0.13913E-01	0.4734E-01	0.2939
t*q	-0.48119E-02	0.4734E-01	-0.1016
m*q	0.31802E-01	0.4734E-01	0.6718
t*m*q	0.29648E-01	0.4734E-01	0.6263
CONSTANT	-0.30473E-01	0.6725E-01	-0.4531

Output L.4

Logistic regression. Previous non-users of M. BMDPLR

STEP NUMBER   0

LOG LIKELIHOOD =   -358.017

TERM	COEFFICIENT	STANDARD ERROR	COEFF/S.E.
temp	-0.27228E-01	0.9292E-01	-0.2930
ls	-0.14350	0.1123	-1.277
t*l	0.10996E-01	0.1123	0.9788E-01
qs	-0.17890E-01	0.6654E-01	-0.2688
t*q	-0.34459E-01	0.6654E-01	-0.5178
CONSTANT	0.28354	0.9292E-01	3.051

A similar table is output for users of M.

Example M

Fertilizer experiment on growth of cauliflowers

Description of data

In an experiment on the effect of nitrogen and potassium upon the growth of cauliflowers, four levels of nitrogen and two levels of potassium were tested:

Nitrogen levels: 0, 60, 120, 180 units per acre (coded as 0, 1, 2, 3);
Potassium levels: 200, 300 units per acre (coded as A, B).

The experiment was arranged in 4 blocks, each containing 4 plots as shown below. When harvested, the cauliflowers were graded according to size. Table M.1 shows the yield (number of cauliflowers) of different sizes: grade 12, for example, means that 12 cauliflowers fit into a standard size crate. The data were provided by Mr J.C. Gower, Rothamsted Experimental Station.

Table M.1. Numbers of cauliflowers of each grade

Block	Treatment	Grade				Unmarketable
		12	16	24	30	
I	0A	—	1	21	24	2
	2B	1	6	24	13	4
	1B	—	4	28	12	4
	3A	1	10	26	9	1
II	3B	—	4	26	14	4
	1A	—	5	27	13	3
	0B	—	—	12	28	8
	2A	—	5	35	5	3
III	1B	—	1	22	22	3
	0A	—	1	8	33	3
	3A	—	6	22	17	2
	2B	—	3	27	14	4
IV	0B	—	—	8	30	10
	2A	—	7	16	22	3
	3B	—	2	31	11	4
	1A	—	—	13	26	9

The analysis

The design of the experiment is a complete 2×4 factorial, confounded into two blocks I, II, with blocks III, IV forming a replicate of the design. The interaction of potassium with the quadratic component of nitrogen, i.e. $K \times N_Q$, defines the confounding.

The problem reduces to one of estimating orthogonal single degree of freedom contrasts. The three contrasts representing $R(=$ replicate$)$, $K \times N_Q$ and $R \times K \times N_Q$ correspond to the 3 d.f. between blocks and hence are not of direct interest, although they need to be fitted in order to obtain a correct residual mean square. The effects of prime interest are main effects K, $N_L(=$ linear nitrogen$)$, N_Q and the interaction $K \times N_L$. This leaves 8 d.f. for error.

The linear and quadratic coefficients for nitrogen $(-3, -1, 1, 3$ and $1, -1, -1, 1)$ are easily generated in a **transform** paragraph.

The response measure Y analysed in *App. Stat.* is the effective number of crates of marketable cauliflowers, given by

$$Y = \tfrac{1}{12}n_{12} + \tfrac{1}{16}n_{16} + \tfrac{1}{24}n_{24} + \tfrac{1}{30}n_{30}. \qquad \text{(M.1, App. Stat.)}$$

The analysis with Y as dependent variable is easily done by multiple regression.

An alternative response measure Z, the proportion of cauliflowers of grade 24 or better, given by

$$Z = (n_{12} + n_{16} + n_{24})/48 \qquad \text{(M.2, App. Stat.)}$$

is not analysed in *App. Stat.*

Instructions are given for analysing both the above response variables, Y by ordinary multiple regression and Z by logistic regression. Note, however, that the 'usual' formulae for standard errors will probably underestimate the true 'error' because of positive correlation between the qualities of cauliflowers in a plot.

Programs

BMDP1R is used for a multiple regression of Y on the seven orthogonal contrasts K, N_L, N_Q, $K \times N_L$, R, $K \times N_Q$, $R \times K \times N_Q$. The instructions are:

```
/problem     title = 'example m.  cauliflowers.
             bmdp1r.'.
/input       variables = 7.  format = free.
/variable    names  =  rep, nitrgen, potsm, n12,
             n16, n24, n30.  add = new.
/transform   y = n12/12 + n16/16 + n24/24 +        response Y
             n30/30.
             r = 2*rep - 3.
             k = 2*potsm - 3.                      coefficients of 7 orthogonal con-
                                                   trasts generated
```

```
           nl = 2*nitrgen - 3.
           nq = (nl**2 - 5)/4.
           kxnl = k*nl.
           kxnq = k*nq.
           rxkxnq = r*k*nq.
/regress   dependent = y.
           independent = k, nl, nq, kxnl, r,
           kxnq, rxkxnq.
/print     data.
```
print data, predicted values and
residuals
```
/plot      residual. normal.
```
plot residuals v. predicted
normal probability plot
```
/end
1 0 1 0 1 21 24
1 2 2 1 6 24 13

      . . .
```
16 lines of data
Code: Blocks I, II = 1
III, IV = 2
Potassium A = 1
B = 2

The **print** and **plot** paragraphs are optional. Plots of residuals against any of the variables may also be requested by adding to the **plot** paragraph the sentence

variables =   .

filling in the list of names (any from the **names** list in the **variable** paragraph).

Output M.1 shows the residual m.s. from regression, the estimated coefficients and s.e.s. This is equivalent to the analysis of variance of Table M.4, *App. Stat*. The significance of N_L is apparent, as is the suggestion of $K \times N_Q$, and to a lesser extent K and N_Q. The estimated yield at each of the four levels of N (Table M.5, *App. Stat*.) can be obtained by averaging the appropriate predicted values given in the output (not shown here).

Note: the instructions given above for BMDP1R may be used instead with BMDP2R, replacing the **regress** paragraph by

/regress   **dependent = y.**
           independent = a,k,nl,nq,kxnl.
           setn = a.
           a = r,kxnq,rxkxnq.
           force = 1.

where **force = 1.** ensures that all variables enter (in turn) and the three contrasts R, $K \times N_Q$ and $R \times K \times N_Q$ (= 3 d.f. between blocks) enter as one set.

For a logistic regression of Z on the same seven contrasts, we use BMDPLR. This is a stepwise program but by specifying **model** we can ensure that all seven contrasts are included. The instructions (many of which are the same as for BMDP1R) are as follows:

```
/problem   title = 'example m. cauliflowers.
           bmdplr'.
/input     variables = 7. format = free.
/variable  names = rep, nitrgen, potsm, n12,
           n16, n24, n30.  add = new.
           use = over24 to rxkxnq.
```
use only the generated variables

```
/transform    over24 = n12 + n16 + n24.
              total = 48.
              r = 2*rep - 3.                    generate Z numerator, Z denomi-
              k = 2*potsm - 3.                  nator, and coefficients of 7 con-
              nl = 2*nitrgen - 3.               trasts
              nq = (nl**2 - 5)/4.
              kxnl = k*nl.
              kxnq = k*nq.
              rxkxnq = r*k*nq.
/regress      count = total.
              scount = over24.                  Z = scount/total
              model = k, nl, kxnl, nq, kxnq,    all contrasts included
              r, rxkxnq.
              interval = k, nl, nq, kxnl, kxnq, variables assumed categorical
              rxkxnq.                           unless stated as 'interval'
/print        cells = use.
/end
1 0 1 0 1 21 24                                 16 lines of data as before
1 2 2 1 6 24 13
    . . .
```

Output M.2 shows the estimated coefficients of the logistic model and goodness-of-fit statistics. The conclusions are broadly similar to those from the analysis of derived response Y although the goodness-of-fit statistics indicate some departure from the model, which could be due to positive correlation among errors as noted above. Note also from Output M.2 that the coefficients are no longer completely orthogonal, although almost so. Output M.3 shows fitted probabilities and residuals. The standardized residuals for Block III, 2B (-2.7805) and Block IV, 3B (3.4325) indicate that further examination of the model is desirable.

Suggestions for further work

(i) Examine in more detail the adequacy of fit of the above logistic model and fit any further logistic models which might be helpful in formulating your conclusions to be drawn from the analysis.

(ii) Analyse Z as a quantitative response using BMDP1R.

(iii) Compare critically the answers from all the above analyses.

Output M.1

Derived response Y. Multiple regression. BMDP1R

```
ANALYSIS OF VARIANCE
                      SUM OF SQUARES    DF    MEAN SQUARE    F RATIO   P(TAIL)
          REGRESSION       0.5823       7       0.0832        7.694    0.0050
          RESIDUAL         0.0865       8       0.0108
```

VARIABLE		COEFFICIENT	STD. ERROR	STD. REG COEFF	T	P(2 TAIL)	TOLERANCE
INTERCEPT		1.73672					
r	9	-0.08568	0.02600	-0.419	-3.296	0.0109	1.00000
k	10	-0.04297	0.02600	-0.210	-1.653	0.1369	1.00000
nl	11	0.06547	0.01163	0.716	5.631	0.0005	1.00000
nq	12	-0.04505	0.02600	-0.220	-1.733	0.1213	1.00000
kxnl	13	-0.67709E-03	0.11626E-01	-0.007	-0.058	0.9550	1.00000
kxnq	14	-0.06120	0.02600	-0.299	-2.354	0.0464	1.00000
rxkxnq	15	0.78122E-03	0.25996E-01	0.004	0.030	0.9768	1.00000

Note that R, $K \times N_Q$ and $R \times K \times N_Q$ correspond to 3 d.f. between blocks.

Output M.2

Derived response Z. Logistic regression. BMDPLR

```
STEP NUMBER   0
---------------
```

```
                                LOG LIKELIHOOD =  -475.381
GOODNESS OF FIT CHI-SQ   (2*0*LN(O/E)) =   20.614   D.F.=   8   P-VALUE= 0.008
GOODNESS OF FIT CHI-SQ   ( D. HOSMER ) =    9.921   D.F.=   8   P-VALUE= 0.271
GOODNESS OF FIT CHI-SQ   ( C.C.BROWN ) =    6.646   D.F.=   2   P-VALUE= 0.036
```

TERM	COEFFICIENT	STANDARD ERROR	COEFF/S.E.
k	-0.41701E-01	0.7826E-01	-0.5329
nl	0.29567	0.3619E-01	8.170
kxnl	0.24328E-01	0.3608E-01	0.6743
nq	-0.26562	0.7838E-01	-3.389
kxnq	-0.13162	0.7831E-01	-1.681
r	-0.41049	0.7854E-01	-5.227
rxkxnq	-0.33331E-01	0.7830E-01	-0.4257
CONSTANT	0.92295E-01	0.7820E-01	1.180

```
CORRELATION MATRIX OF COEFFICIENTS
----------------------------------
```

	k	nl	kxnl	nq	kxnq	r	rxkxnq	CONSTANT
k	1.000							
nl	-0.046	1.000						
kxnl	-0.042	0.045	1.000					
nq	0.020	-0.065	-0.054	1.000				
kxnq	0.068	-0.054	-0.058	0.031	1.000			
r	0.009	-0.120	-0.016	0.049	0.020	1.000		
rxkxnq	0.045	-0.015	-0.092	0.009	-0.018	0.014	1.000	
CONSTANT	0.028	-0.035	-0.046	0.068	0.019	-0.019	0.018	1.000

Output M.3

Derived response Z. Observed and predicted proportions based on logistic regression. BMDPLR

SUMMARY DESCRIPTION OF CELLS.
CELLS ARE FORMED BY ALL COMBINATIONS OF VALUES OF ALL VARIABLES.

NUMBER over24	NUMBER FAILURE	OBSERVED PROPORTION over24	PREDICTED PROB.OF over24	S.E. OF PREDICTED PROB.	OBS-PRED S.E.RES.	PRED. LOG ODDS	CHI	DEVIANCE	HAT MATRIX DIAGONAL	INFLUENCE
8	40	0.1667	0.1480	0.0349	0.4980	-1.750	0.3644	0.3584	0.4645	0.2151
9	39	0.1875	0.2330	0.0447	-1.0973	-1.191	-0.7464	-0.7652	0.5373	1.3980
12	36	0.2500	0.2967	0.0535	-1.2121	-0.863	-0.7090	-0.7209	0.6579	2.8252
13	35	0.2708	0.3900	0.0453	-2.2101	-0.447	-1.6932	-1.7355	0.4131	3.4385
22	26	0.4583	0.3925	0.0566	1.5702	-0.437	0.9343	0.9266	0.6460	4.4986
23	25	0.4792	0.5239	0.0474	-0.8236	0.096	-0.6200	-0.6196	0.4332	0.5184
23	25	0.4792	0.4380	0.0466	0.7678	-0.249	0.5750	0.5733	0.4243	0.4232
28	20	0.5833	0.6075	0.0566	-0.5766	0.437	-0.3431	-0.3419	0.6460	0.6067

Design matrix values (k / nq / rxkxnq, nq / kxnq, nl / kxnq, kxnl / r):

k nq rxkxnq			nq kxnq		nl kxnq		kxnl r	
1.00	1.00	1.00	1.00	1.00	-3.00	1.00	-3.00	1.0
-1.00	1.00	-1.00	1.00	-1.00	-3.00	-1.00	3.00	1.0
1.00	1.00	1.00	-1.00	-1.00	-3.00	1.00	-3.00	-1.0
-1.00	1.00	-1.00	1.00	1.00	-1.00	1.00	1.00	1.0
-1.00	1.00	1.00	-1.00	-1.00	-3.00	1.00	3.00	-1.0
-1.00	1.00	-1.00	1.00	1.00	1.00	1.00	-1.00	1.0
1.00	-1.00	-1.00	1.00	-1.00	-1.00	-1.00	-1.00	1.0
-1.00	1.00	-1.00	1.00	1.00	3.00	-1.00	-3.00	1.0

Output M.3 (continued)

SUMMARY DESCRIPTION OF CELLS.
CELLS ARE FORMED BY ALL COMBINATIONS OF VALUES OF ALL VARIABLES.

NUMBER over24	NUMBER FAILURE	OBSERVED PROPORTION over24	PREDICTED PROB.OF over24	S.E. OF PREDICTED PROB.	OBS-PRED S.E.RES.	PRED. LOG ODDS	CHI	DEVIANCE	HAT MATRIX DIAGONAL	INFLUENCE	k / nq / rxkxnq	nl / kxnq	kxnl / r
30	18	0.6250	0.5964	0.0459	0.5292	0.391	0.4032	0.4049	0.4194	0.2023	1.00 / -1.00 / -1.00	1.00 / -1.00	1.00 / 1.0
30	18	0.6250	0.7421	0.0470	-2.7805	1.057	-1.8554	-1.7820	0.5547	9.6319	1.00 / 1.00 / -1.00	3.00 / 1.00	3.00 / -1.0
31	17	0.6458	0.7586	0.0368	-2.2725	1.145	-1.8256	-1.7477	0.3546	2.8376	1.00 / -1.00 / 1.00	1.00 / -1.00	1.00 / -1.0
32	16	0.6667	0.6084	0.0468	1.1073	0.440	0.8275	0.8360	0.4416	0.9699	-1.00 / -1.00 / -1.00	-1.00 / 1.00	1.00 / -1.0
32	16	0.6667	0.6236	0.0465	0.8245	0.505	0.6154	0.6206	0.4428	0.5403	-1.00 / -1.00 / 1.00	-1.00 / -1.00	-1.00 / -1.0
33	15	0.6875	0.5423	0.0581	3.4325	0.170	2.0194	2.0520	0.6539	22.2590	1.00 / 1.00 / 1.00	3.00 / 1.00	3.00 / 1.0
37	11	0.7708	0.7670	0.0447	0.0935	1.191	0.0636	0.0637	0.5373	0.0101	-1.00 / 1.00 / 1.00	3.00 / -1.00	-3.00 / -1.0
40	8	0.8333	0.7277	0.0393	2.0764	0.983	1.6436	1.7284	0.3734	2.5696	-1.00 / -1.00 / -1.00	1.00 / 1.00	-1.00 / -1.0

Cells are listed in order of 'observed proportion over 24'.

Example N   Subjective preference data on soap pads

Description of data

Table N.1 gives data obtained during the development of a soap pad. The factors, amount of detergent, d, coarseness of pad, c, and solubility of detergent, s, were each set at two levels. There were 32 judges and the experiment was done on two days. Each judge attached a score (excellent $= 1, \ldots,$ poor $= 5$) to two differently formulated pads on each of two days. For the data and several different analyses, see Johnson (1967).

The analysis

The experimental design is rather complex, and is discussed in *App. Stat.*

Preliminary inspection of the data involving calculation of marginal one- or two-way means, or tabulation of the frequencies of scores, is straightforward. We use BMDP9D and BMDP4F.

For the main analysis we follow the approach in *App. Stat.* and ignore any repetition of judges, i.e. we compute the analysis assuming there to be 32 different pairs of judges. Inspection of residuals can be used as a check on the justification of this. Scores on the five-point scale are treated as an ordinary quantitative variable; note that there are not many extreme values (1 or 5).

The main analysis of variance table, Table N.4, *App. Stat.*, is reproduced here as Table N.2. In principle, this can be obtained by treating the design as a complete factorial with 2^3 treatments (T) $\times 2$ days (D) $\times 2$ judges (J) $\times 4$ replicates (R), pooling together certain of the sums of squares and degrees of freedom to give the following terms:

	Pooled s.s.	Total d.f.
Between judges, within treatments, within replicates	J, T $\times$ J, J $\times$ R, T $\times$ J $\times$ R	32
Days $\times$ between judges	D $\times$ J, T $\times$ D $\times$ J, D $\times$ J $\times$ R, T $\times$ D $\times$ J $\times$ R	32

It may be necessary to increase the program capacity to cope with a factorial analysis of this size.

85

Table N.1. Subjective scores allocated to soap pads prepared in accordance with 2^3 factorial scheme. Five-point scale; 1 = excellent, 5 = poor

Judge	Treatment	Day 1	Day 2	Judge	Treatment	Day 1	Day 2
Replicate I				Replicate II			
1	1	2	4	5	1	4	2
17	1	2	3	21	1	3	3
1	dcs	4	4	5	cs	3	4
17	dcs	4	4	21	cs	1	2
2	d	5	4	6	d	1	2
18	d	4	4	22	d	5	4
2	cs	2	1	6	dcs	3	3
18	cs	1	2	22	dcs	4	4
3	c	1	3	7	c	3	3
19	c	5	5	23	c	3	5
3	ds	3	2	7	s	4	4
19	ds	4	3	23	s	5	3
4	s	1	3	8	dc	4	4
20	s	2	3	24	dc	2	3
4	dc	3	4	8	ds	3	2
20	dc	3	3	24	ds	2	3
Replicate III				Replicate IV			
9	1	3	2	13	1	3	4
25	1	2	3	29	1	3	4
9	ds	4	3	13	dc	2	3
25	ds	3	3	29	dc	3	4
10	d	1	1	14	d	4	4
26	d	3	3	30	d	4	3
10	s	2	1	14	c	2	2
26	s	1	1	30	c	4	5
11	c	3	3	15	s	5	5
27	c	3	3	31	s	3	3
11	dcs	3	3	15	dcs	4	4
27	dcs	2	2	31	dcs	1	2
12	dc	3	3	16	ds	4	3
28	dc	4	4	32	ds	4	3
12	cs	1	2	16	cs	3	4
28	cs	3	3	32	cs	1	4

Table N.2. Analysis of variance ignoring repetition of judges
(Table N.4, App. Stat.)

	d.f.	s.s.	m.s.
s	1	3.4453	
c	1	0.0078	
d	1	3.4453	
c × d	1	0.1953	
d × s	1	1.3203	
s × c	1	1.3203	
s × c × d	1	2.2578	
Treatments	7	11.9922	1.713
Replicates	3	11.6484	3.883
Treatments × Replicates	21	38.0394	1.811
Days	1	1.3203	1.320
Days × Replicates	3	1.3983	0.466
Days × Treatments	7	6.7422	0.963
Days × Treatments × Replicates	21	10.2891	0.490
Between judges, within treatments, within replicates	32	55.7500	1.742
Days × Between judges	32	11.7500	0.367
Total	127		

Alternatively, we can first analyse each of the four replicates separately and inspect for consistency before combining the results. For each replicate of the 2^3 treatments × 2 days × 2 judges factorial, we pool together s.s and d.f. to give the following terms:

	Pooled s.s.	Total d.f.
Between judges, within treatments	J, T × J	8
Judges × days within treatments	D × J, T × D × J	8

These are then pooled over the four replicates to give the earlier terms each with 32 d.f. This is followed by an analysis for the 2^3 treatments × 2 days × 4 replicates factorial and an analysis for the simple 2^3 treatment factorial. In this way all the lines of Table N.4, App. Stat. are obtained. This is the procedure adopted here.

Program

BMDP9D is used to obtain one- and two-way marginal means for days and replicates. The instructions are:

```
/problem        title = 'example n.  soap pads.
                bmdp9d.'.
/input          variables = 7.  format = free.
/variable       names = replicate, judge, d, c, s,      y = score
                day, y.
/tabulate       grouping = day, replicate.
                variable = y.
                margin = 'dr', 'd.', '.r'.              1- and 2-way margins
/end
1 1 0 0 0 1 2                                           128 lines of data
1 1 0 0 0 2 4
                                               Code: treatments: 1 = 0 0 0
                                                                 d = 1 0 0
     . . .                                                      cs = 0 1 1
                                                                    etc.
```

Output N.1 shows the two-way marginal means for days × replicates (as in Table N.2, *App. Stat.*).

BMDP4F is used to obtain the frequency distribution of scores for each treatment. The instructions are:

```
/problem        title = 'example n.  soap pads.
                bmdp4f.'.
/input          variables = 7.  format = free.
/variable       names = replicate, judge, d, c, s,
                day, y. add = new.
/group          codes(treatmt) = 0 to 7.
                names(treatmt) = '1', s, c, sc, d,      labels for output
                ds, dc, dsc.
/transform      treatmt = 4*d + 2*c + s.                compound code for treatments
/table          columns = treatmt.
                rows = y.
/end
1 1 0 0 0 1 2                                           128 lines of data as before
1 1 0 0 0 2 4


     . . .
```

Output N.2 shows the frequency distributions of scores (as in Table N.3, *App. Stat.*).

Marginal mean scores (as in the last row of Table N.3, *App. Stat.*) for each treatment can be obtained using BMDP9D, simply replacing the **table** paragraph in the above instructions (given for BMDP4F) by

/tabulate   grouping = treatment.
             variable = y.

BMDP2V is used for analysis of variance of the factorial arrangements explained earlier. The instructions for replicate 1 of the 2^3 treatments × 2 days × 2 judges factorial, pooling J and T × J into the error term, are:

```
/problem        title = 'example n.  soap pads.
                txdxj. bmdp2v.'.
/input          variables = 7.  format = free.
/variable       names = replicate, judge, d, c, s,
                day, y. add = new.
/transform      treatmt = 4*d + 2*c + s.
                use = replicate eq 1.
```

```
/design      dependent = y.
             grouping = judge, day, treatmt.
             exclude = 1, 13.                        model excludes J and T × J
/end
1 1 0 0 0 1 2                                         128 lines of data as before
1 1 0 0 0 2 4

  . . .
```

For replicate 2, amend the **use** statement in the **transform** paragraph to be

use = replicate eq 2.

and similarly for replicates 3, 4. (If wished, all four replicates may be analysed in one submission of the data by using the **for–%** feature. See note 9(ii) in the introductory section, p. 9.) Output N.3 shows the analysis of variance for each replicate. Note that the error term represents 'between judges, within treatments'; $J \times D$ and $J \times D \times T$ need to be pooled to give 'judges × days within treatments' with 8 d.f. When pooled over the four replicates these terms give the s.s. for the two lines with 32 d.f. in Table N.4, *App. Stat.*

The remaining s.s. and d.f. in the lower part of Table N.4, *App. Stat.* are given by an analysis of the 2^3 treatments $\times$ 2 days $\times$ 4 replicates using BMDP2V. The instructions are:

```
/problem     title = 'example n.  soap pads.
             txdxr. bmdp2v.'.
/input       variables = 7.  format = free.
/variable    names = replicate, judge, d, c, s,
             day, y.  add = new.
/transform   treatmt = 4*d + 2*c + s.
/design      dependent = y.
             grouping = replicate, day, treatmt.
/end
1 1 0 0 0 1 2                                         128 lines of data as before
1 1 0 0 0 2 4

  . . .
```

Output N.4 shows the analysis of variance.

A final analysis of variance for the simple 2^3 treatment factorial $D \times C \times S$ gives the **s.s.** and **d.f.** for the upper part of Table N.4, *App. Stat.* The instructions for BMDP2V are:

```
/problem     title = 'example n.  soap pads.
             dxcxs. bmdp2v.'.
/input       variables = 7.  format = free.
/variable    names = replicate, judge, d, c, s,
                     day, y.
/design      dependent = y.
             grouping = d, c, s.
/end
1 1 0 0 0 1 2                                         128 lines of data as before
1 1 0 0 0 2 4

  . . .
```

Output N.5 shows the analysis of variance.

Other programs

BMDP4V and BMDP8V can be used for analysis of variance of factorial designs.

Suggestions for further work

(i) Print and inspect residuals for the above analysis. To print residuals and fitted values when using BMDP2V add the statement

print.

to the **design** paragraph.

(ii) Define a response variable Z,

$$Z = \begin{cases} 0 & \text{if score less than } k \\ 1 & \text{otherwise} \end{cases}$$

where $1 < k \leqslant 5$, and use BMDPLR to fit a logistic model.

(iii) Compare your conclusions from (ii) with those from the analysis of the straight score.

(iv) Consider how the logistic models of (ii) for $k = 2, \ldots, 5$ may be combined.

Output N.1

Two-way marginal means for days × replicates. BMDP9D

```
CELL      day        replicat
NUMBER      6            1
   1      *1          *1
   2      *2          *1
   3      *1          *2
   4      *2          *2
   5      *1          *3
   6      *2          *3
   7      *1          *4
   8      *2          *4
```

DESCRIPTIVE STATISTICS FOR NON-EMPTY CELLS

```
CELL       VARIABLE   7 y
NUMBER    FREQ.    MEAN    STD.DEV.

   1      16.     2.87500   1.3601
   2      16.     3.25000   1.0000
   ----------------------------
   3      16.     3.12500   1.2042
   4      16.     3.18750   0.9106
   ----------------------------
   5      16.     2.56250   0.9639
   6      16.     2.50000   0.8944
   ----------------------------
   7      16.     3.12500   1.1475
   8      16.     3.56250   0.8921
```

Output N.2

Frequency distributions of response versus treatment BMDP4F

***** OBSERVED FREQUENCY TABLE 1

y				treatmt				
	1	s	c	sc	d	ds	dc	dsc	TOTAL
ONE	0	4	1	5	3	0	0	1	14
TWO	5	2	2	4	1	3	2	3	22
THREE	7	5	8	4	3	9	8	4	48
FOUR	4	2	1	3	7	4	6	8	35
FIVE	0	3	4	0	2	0	0	0	9
TOTAL	16	16	16	16	16	16	16	16	128

Output N.3

Analysis of variance for 2^3 treatments $\times$ 2 days $\times$ 2 judges. BMDP2V

(a) Replicate 1

SOURCE	SUM OF SQUARES	DEGREES OF FREEDOM	MEAN SQUARE	F	TAIL PROB.
MEAN	300.12500	1	300.12500	218.27	0.0000
day	1.12500	1	1.12500	0.82	0.3921
treatmt	22.87500	7	3.26786	2.38	0.1242
jd	0.12500	1	0.12500	0.09	0.7707
dt	5.87500	7	0.83929	0.61	0.7354
jdt	2.87500	7	0.41071	0.30	0.9355
ERROR	11.00000	8	1.37500		

(b) Replicate 2

SOURCE	SUM OF SQUARES	DEGREES OF FREEDOM	MEAN SQUARE	F	TAIL PROB.
MEAN	318.78125	1	318.78125	147.84	0.0000
day	0.03125	1	0.03125	0.01	0.9071
treatmt	7.46875	7	1.06696	0.49	0.8153
jd	0.28125	1	0.28125	0.13	0.7273
dt	4.21875	7	0.60268	0.28	0.9449
jdt	4.96875	7	0.70982	0.33	0.9195
ERROR	17.25000	8	2.15625		

(c) Replicate 3

SOURCE	SUM OF SQUARES	DEGREES OF FREEDOM	MEAN SQUARE	F	TAIL PROB.
MEAN	205.03125	1	205.03125	187.46	0.0000
day	0.03125	1	0.03125	0.03	0.8700
treatmt	14.71875	7	2.10268	1.92	0.1897
jd	0.28125	1	0.28125	0.26	0.6258
dt	0.71875	7	0.10268	0.09	0.9973
jdt	1.46875	7	0.20982	0.19	0.9788
ERROR	8.75000	8	1.09375		

(d) Replicate 4

SOURCE	SUM OF SQUARES	DEGREES OF FREEDOM	MEAN SQUARE	F	TAIL PROB.
MEAN	357.78125	1	357.78125	152.65	0.0000
day	1.53125	1	1.53125	0.65	0.4423
treatmt	4.96875	7	0.70982	0.30	0.9334
jd	0.28125	1	0.28125	0.12	0.7380
dt	6.21875	7	0.88839	0.38	0.8906
jdt	1.46875	7	0.20982	0.09	0.9977
ERROR	18.75000	8	2.34375		

Output N.4

Analysis of variance for 2^3 treatments × 2 days × 4 replicates. BMDP2V

SOURCE	SUM OF SQUARES	DEGREES OF FREEDOM	MEAN SQUARE	F	TAIL PROB.
MEAN	1170.07031	1	1170.07031	1109.40	0.0000
replicat	11.64844	3	3.88281	3.68	0.0164
day	1.32031	1	1.32031	1.25	0.2674
treatmt	11.99219	7	1.71317	1.62	0.1446
rd	1.39844	3	0.46615	0.44	0.7238
rt	38.03906	21	1.81138	1.72	0.0511
dt	6.74219	7	0.96317	0.91	0.5022
rdt	10.28906	21	0.48996	0.46	0.9736
ERROR	67.50000	64	1.05469		

Output N.5

Analysis of variance for 2^3 treatments, D × C × S. BMDP2V

SOURCE	SUM OF SQUARES	DEGREES OF FREEDOM	MEAN SQUARE	F	TAIL PROB.
MEAN	1170.07031	1	1170.07031	1025.35	0.0000
d	3.44531	1	3.44531	3.02	0.0849
c	0.00781	1	0.00781	0.01	0.9342
s	3.44531	1	3.44531	3.02	0.0849
dc	0.19531	1	0.19531	0.17	0.6798
ds	1.32031	1	1.32031	1.16	0.2842
cs	1.32031	1	1.32031	1.16	0.2842
dcs	2.25781	1	2.25781	1.98	0.1621
ERROR	136.93750	120	1.14115		

Example O

Atomic weight of iodine

Description of data

Table O.1 gives ratios of reacting weight of iodine and silver obtained, in an accurate determination of atomic weight of iodine, using five batches of silver A, B, C, D, E and two of iodine, I, II (Baxter and Landstredt, 1940; Brownlee, 1965). Silver batch C is a repurification of batch B, which in turn is a repurification of batch A. In these data 1.176 399 has been subtracted from all values.

Table O.1. Ratios of reacting weight with 1.176 399 subtracted $\times 10^6$

Silver batch	Iodine batch	
	I	*II*
A	23, 26	0, 41, 19
B	42, 42	24, 14
C	30, 21, 38	
D	50, 51	62
E	56	

The analysis

This example illustrates in very simple form aspects arising widely in the analysis of unbalanced data. Because of the lack of balance it is necessary to proceed by fitting a sequence of models and differencing the residual sums of squares from two models to obtain the appropriate sum of squares for testing columns (adjusted for rows) or rows (adjusted for columns). If Y_{ijk} denotes the kth observation in row i and column j, the sequence of models suggested is:

Model I $E(Y_{ijk}) = \mu$; homogeneity
Model II$_1$ $E(Y_{ijk}) = \mu + \alpha_i$; pure row effects
Model II$_2$ $E(Y_{ijk}) = \mu + \beta_j$; pure column effects

94

Model II_{12} $E(Y_{ijk}) = \mu + \alpha_i + \beta_j$; additivity (no interaction)

Model III $E(Y_{ijk}) = \mu_{ij} = \mu + \alpha_i + \beta_j + \gamma_{ij}$; arbitrary means.

(O.1, App. Stat.)

Were it not for some empty cells in the two-way table, BMDP2V would fit this sequence in one run of the program. The empty cells necessitate a separate run for Model III.

Program

BMDP2V is used. To run the program, for all but Model III, we substitute fictitious values (e.g. zero) to represent observed values in the empty cells and assign zero weight to these cells. We assign unit weight to all nonempty cells. The instructions are:

```
/problem     title = 'example o.
             iodine atomic weight.  bmdp2v.'.
/input       variables = 4. format = free.
/variable    names = silver, iodine, y, w.         y = observed value, w = weight
             weight = w.
/group       codes(silver) = 1 to 5.
             names(silver) = a, b, c, d, e.
             codes(iodine) = 1, 2.                 names for output
             names(iodine) = batch1, batch2.
/design      dependent = y.
             grouping = silver, iodine.
             included = 1, 2.                       main effects only in model
             print.                                 print predicted values, residuals
/end
1 1 23 1                                            18 lines of data
1 1 26 1                                            Code: col 1. silver A = 1, B = 2, ..., E = 5
                                                           col 4. w = 0 (empty cell), 1 (non-
     ...                                                    empty)
```

The data for an empty cell, for example, for silver c, iodine II would read 3 2 0 0. Output O.1 shows the analysis of variance table. The s.s. for silver (adjusted for iodine) is the difference in residual s.s. from models II_2 and II_{12}, that for iodine (adjusted for silver) the difference between II_1 and II_{12}. Output O.2 shows the predicted values and residuals based on model II_{12}; note the empty cells correspond to cases numbered 13 and 18. Averaging the predicted values over iodine batches gives the estimated silver batch means quoted in Table O.3(a), *App. Stat.*

Due to the empty cells there are too many parameters for model III to be fitted by instructions similar to those given above. Instead we regard the data as a one-way classification of (nonempty) cells. Keeping the input data in the same format we simply add a **transform** paragraph to generate cell codes 1, ..., 10 and use nonempty cells for a between- and within-cells analysis of variance. The **group** paragraph is not required (although it need not necessarily be removed). The instructions, again using BMDP2V, are:

```
/problem      title = 'example o.  iodine.
              model III.  bmdp2v.'.
/input        variables = 4.  format = free.
/variable     names = silver, iodine, y, w.
              weight = w.
              add = new.
/transform    cell = 2*(silver -1) + iodine.     cell codes 1, ... , 10
              use = w gt 0.                       use nonempty cells
/design       dependent = y.
              grouping = cell.
/end
1 1 23 1                                          18 lines of data as before
1 1 26 1

   . . .
```

Output O.3 shows the analysis of variance. The difference in error s.s. in Output O.1 and O.3 gives the s.s. with 2 d.f. for iodine × silver; the error m.s. of Output O.3 is used for assessing the significance of iodine × silver, iodine (adjusted for silver) and silver (adjusted for iodine); see Table O.2(b), *App. Stat.*

Suggested further work

Examine the residuals of Output O.2 and carry out any further analysis that might be indicated.

Output O.1

Analysis of variance based on models I, II_1, II_2 and II_{12}. BMDP2V

SOURCE	SUM OF SQUARES	DEGREES OF FREEDOM	MEAN SQUARE	F	TAIL PROB.
MEAN	13555.13396	1	13555.13396	88.41	0.0000
silver	2249.05504	4	562.26376	3.67	0.0435
iodine	149.95504	1	149.95504	0.98	0.3460
ERROR	1533.17829	10	153.31783		

In this table the sums of squares are for silver (adjusted for iodine) and iodine (adjusted for silver).

Output O.2

Predicted values and residuals based on model II_{12}. BMDP2V

CASE	silver	iodine	PREDICTD	RESIDUAL
1	a	batch1	26.13953	-3.13953
2	a	batch1	26.13953	-0.13953
3	a	batch2	18.90698	-18.90698
4	a	batch2	18.90698	22.09302
5	a	batch2	18.90698	0.09302
6	b	batch1	34.11628	7.88372
7	b	batch1	34.11628	7.88372
8	b	batch2	26.88372	-2.88372
9	b	batch2	26.88372	-12.88372
10	c	batch1	29.66667	0.33333
11	c	batch1	29.66667	-8.66667
12	c	batch1	29.66667	8.33333
13	c	batch2	22.43411	-22.43411
14	d	batch1	56.74419	-6.74419
15	d	batch1	56.74419	-5.74419
16	d	batch2	49.51163	12.48837
17	e	batch1	56.00000	0.00000
18	e	batch2	48.76744	-48.76744

Output O.3

Analysis of variance based on model III. BMDP2V

SOURCE	SUM OF SQUARES	DEGREES OF FREEDOM	MEAN SQUARE	F	TAIL PROB.
MEAN	19760.02381	1	19760.02381	151.76	0.0000
cell	3213.77083	7	459.11012	3.53	0.0491
ERROR	1041.66667	8	130.20833		

Example P

Multifactor experiment on a nutritive medium

Description of data

Fedorov, Maximov and Bogorov (1968) obtained the data in Table P.1 from an experiment on the composition of a nutritive medium for green sulphur bacteria *Chlorobrium thiosulphatophilum*. The bacteria were grown under constant illumination at a temperature of 25–30 °C: the yield was determined during the stationary phase of growth. Each factor was at two levels, with each level used 8 times. Subject to this, the factor levels were randomized.

The analysis

Although this problem raises some difficult statistical issues, computationally it is straightforward, involving a series of multiple regressions, readily fitted by BMDP1R. Alternatively, BMDP9R (all subsets regression) may be used.

The analysis in *App. Stat.* starts by fitting a main-effects model containing all 10 component variables $x_1, \ldots, x_{10}$. The residual mean square from this model is appreciably greater than the external estimate of variance (3.8^2) and so models containing cross-product terms are fitted but, with only 16 observations, the choice of possible terms is restricted. A model is built up from a small number of main effects, choosing those found to be appreciable in the initial analysis, plus their interactions. This leads to a reasonably well fitting model with four main effects, x_6, x_7, x_8, x_9, and two cross-products, $x_6 x_8$ and $x_7 x_8$. There may be other models fitting equally well.

It should be noted that some orthogonality exists between the variables. We here use BMDP1R. Instructions for BMDP9R are given at the end as suggested further work for the reader.

Program

BMDP1R is used. This will fit a specified multiple regression or, by repeating the **regress** paragraph, it will fit a sequence of models. For example, the instructions below fit a regression on x_8, followed by a regression on $x_6, x_7, x_8, x_9, x_6 x_8$ and $x_7 x_8$. **print** and **plot** paragraphs are included; these operate after

Table P.1. Yields of bacteria

	Factors										Y
	x_1 NH_4Cl	x_2 KH_2PO_4	x_3 $MgCl_2$	x_4 $NaCl$	x_5 $CaCl_2$	x_6 $Na_2S\cdot 9H_2O$	x_7 $Na_2S_2O_3$	x_8 $NaHCO_3$	x_9 $FeCl_3$	x_{10} micro-elements	Yield
Levels { +	1500	450	900	1500	350	1500	5000	5000	125	15	
{ −	500	50	100	500	50	500	1000	1000	25	5	
1	−	+	+	+	−	+	−	+	−	+	14.0
2	−	−	+	+	−	+	+	−	−	+	4.0
3	+	−	−	+	+	+	−	−	−	−	7.0
4	−	−	+	−	+	+	−	+	+	+	24.5
5	+	−	+	+	+	+	+	−	−	−	14.5
6	+	−	+	−	+	+	+	+	+	+	71.0
7	−	−	−	−	−	−	−	−	−	−	15.5
8	+	+	−	+	+	−	−	+	+	−	18.0
9	−	+	−	+	−	−	+	−	−	+	17.0
10	+	+	+	+	+	−	−	−	+	−	13.5
11	−	+	+	−	+	−	+	+	+	+	52.0
12	+	+	+	−	−	−	+	+	−	−	48.0
13	+	+	−	−	+	−	+	−	+	−	24.0
14	−	+	−	−	−	+	−	−	+	−	12.0
15	+	−	−	−	−	−	−	+	+	+	13.5
16	−	−	−	+	−	+	+	+	−	+	63.0

All the concentrations are given in mg/l, with the exception of factor 10, whose central level (10 ml of solution of micro-elements per litre of medium) corresponds to 10 times the amount of micro-element in Larsen's medium. The yield has a standard error of 3.8.

each regression and if a large number of regressions are to be calculated it might be better to omit these two paragraphs initially in order to reduce the amount of output and then rerun the program to obtain the full output for any promising models. The instructions are:

```
/problem      title = 'example p.
              biochemical experiment.  bmdp1r.'.
/input        variables = 11.  format = free.
/variable     names = nh4cl, kh2po4, mgcl2, nacl,        input variables
              cacl2, na2s, na2s2o3, nahco3, fecl3,
              micro, y.  add = new.
/transform    x68 = na2s*nahco3.
              x78 = na2s2o3*nahco3.
/regress      dependent = y.
              independent = nahco3.
/regress      dependent = y.
              independent = na2s, na2s2o3, nahco3,
              fecl3, x68, x78.
/print        data.  rreg.                              print data, predicted values, resi-
                                                          duals
                                                        print correlations of coefficients
/plot         normal.                                   normal probability plot
/end
-1 1 1 1 -1 1 -1 1 -1 1 14                              16 lines of data
-1 -1 1 1 -1 1 1 -1 -1 1 4                              Code: plus = 1, minus = −1
              . . .
```

A sequence of models can be fitted by amending the **regress, transform** and **variable** paragraphs in the obvious way.

Output P.1 gives details of the final model selected in *App. Stat.*, which is that of the second **regress** paragraph above; the given output shows the analysis of variance, the estimated regression coefficients and s.e.s (see Tables P.2 and P.3, *App. Stat.*). Output P.2 gives the correlation matrix of estimated coefficients; note the high degree of orthogonality. Output P.3 shows the normal probability plot of residuals. Inspection of the predicted values (not shown here) suggests a tendency for the model to fit better at the higher predicted yields.

Suggested further work

(i) Consider what other models might be fitted. In particular, examine whether the regression of log yield on log concentrations gives a better model. (Add $y = \ln(y)$. to the **transform** paragraph to obtain log yield.)

(ii) Consider and implement graphical methods for choosing the interactions for inclusion.

(iii) Instead of repeatedly using BMDP1R, try using BMDP9R (all subsets regression) to identify well fitting models. Care is needed over which variables, such as interactions, to include in order to avoid too large a set, given there are only 16 observations. Instructions for fitting subsets up to size 6 from the 12 variables used previously (10 main effects, 2 interactions) are as follows:

```
/problem      title = 'example p.
              biochemical experiment.  bmdp9r.'.
/input        variables = 11.  format = free.
/variable     names = nh4cl, kh2po4, mgcl2, nacl,
              cacl2, na2s, na2s2o3, nahco3, fecl3,
              micro, y.  add = new.
/transform    x68 = na2s*nahco3.
              x78 = na2s2o3*nahco3.
/regress      dependent = y.  method = rsq.
              independent = nh4cl to micro, x68, x78.
/print        maxvar = 6.
/end
-1 1 1 1 -1 1 -1 1 -1 1 14                        16 lines of data as before
-1 -1 1 1 -1 1 1 -1 -1 1 4
. . .
```

Detailed information on a particular model can be obtained by an additional run on BMDP9R writing **method = none.** in the **regress** paragraph. This gives a multiple regression on whatever **independent** variables are specified in the **regress** paragraph.

Output P.1

Multiple regression on x_6, x_7, x_8, x_9, $x_6 x_8$ and $x_7 x_8$. BMDP1R

```
MULTIPLE R              0.9866          STD. ERROR OF EST.        4.3704
MULTIPLE R-SQUARE       0.9733
```

ANALYSIS OF VARIANCE

	SUM OF SQUARES	DF	MEAN SQUARE	F RATIO	P(TAIL)
REGRESSION	6272.0791	6	1045.3466	54.728	0.0000
RESIDUAL	171.9056	9	19.1006		

VARIABLE		COEFFICIENT	STD. ERROR	STD. REG COEFF	T	P(2 TAIL)	TOLERANCE
INTERCEPT		25.71875					
na2s	6	1.3462	1.1339	0.07	1.19	0.27	0.9286
na2s2o3	7	11.7837	1.1339	0.59	10.39	0.00	0.9286
nahco3	8	11.4663	1.1339	0.57	10.11	0.00	0.9286
fecl3	9	3.2596	1.2121	0.16	2.69	0.02	0.8125
x68	12	4.5937	1.0926	0.23	4.20	0.00	1.0000
x78	13	9.5312	1.0926	0.47	8.72	0.00	1.0000

Output P.2

Correlation matrix of regression coefficients for model in Output P.1. BMDP1R

		na2s	na2s2o3	nahco3	fecl3	x68	x78
		6	7	8	9	12	13
na2s	6	1.0000					
na2s2o3	7	0.0714	1.0000				
nahco3	8	-0.0714	-0.0714	1.0000			
fecl3	9	0.2673	0.2673	-0.2673	1.0000		
x68	12	-0.0000	-0.0000	-0.0000	-0.0000	1.0000	
x78	13	0.0000	0.0000	0.0000	-0.0000	-0.0000	1.0000

Output P.3

Normal probability plot of residuals for model in Output P.1. BMDP1R

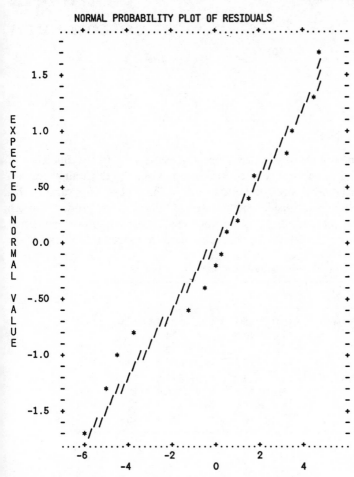

Example Q

Strength of cotton yarn

Description of data

An experiment was done with the objects of estimating: (i) the difference in mean strength of two worsted yarns produced by slightly different processes, and (ii) the variation of strength between and within bobbins for yarns of this type. For each yarn a considerable number of bobbins were produced and 6 bobbins selected at random. From each of these, 4 short lengths were chosen at random for strength testing. The breaking loads are given in Table Q.1.

Table Q.1. Breaking loads (oz)

Bobbin	1	2	3	4	5	6
Yarn A	15.0	15.7	14.8	14.9	13.0	15.9
	17.0	15.6	15.8	14.2	16.2	15.6
	13.8	17.6	18.2	15.0	16.4	15.0
	15.5	17.1	16.0	12.8	14.8	15.5
Yarn B	18.2	17.2	15.2	15.6	19.2	16.2
	16.8	18.5	15.9	16.0	18.0	15.9
	18.1	15.0	14.5	15.2	17.0	14.9
	17.0	16.2	14.2	14.9	16.9	15.5

The analysis

Preliminary analysis involves inspection of means and standard deviations for each of the six bobbins within yarn A, and within yarn B. BMDP7D calculates these, plots a histogram for each bobbin and plots means against standard deviations. BMDP7D also computes one-way analysis of variance so, by separate runs for each yarn and then a combined run, the nested analysis of variance (Table Q.4, *App. Stat.*) can be constructed.

Alternatively, the nested analysis of variance can be obtained using BMDP8V. Yarn is specified as a fixed effect, lengths and bobbins as random effects. The program estimates the components of variance.

Instructions for both BMDP7D and BMDP8V are given.

104

Programs

BMDP7D is first run separately for yarn A, then for yarn B. Instructions for yarn A are given below. For yarn B amend the sentence in the **transform** paragraph to **use = yarn eq 2.**

```
/problem      title = 'example q.  cotton yarn.
              yarna.  bmdp7d'.
/input        variables = 3.  format = free.
/variable     names = yarn, bobbin, brk_load.
/transform    use = yarn eq 1.                        yarn A only
/group        codes(yarn) = 1, 2.
              names(yarn) = yarna, yarnb.             names for output
/histogram    grouping = bobbin.  variable = brk_load.
/print        plot.
/end
1 1 15.0                                              48 lines of data
1 1 17.0
                                            Code: yarn A = 1, B = 2 (col. 1)
   . . .                                          bobbins 1, . . . , 6 (col. 2)
```

Some of the output for yarn A is shown. Outputs Q.1, Q.2 and Q.3 show: Q.1, histogram, mean and standard deviation for each bobbin; Q.2, analysis of variance between and within bobbins; Q.3, diagnostic plots of mean against standard deviation. The slope of the regression of log s.d. on log mean is helpful in checking whether a power transformation is desirable (Box and Cox, 1964). The slope for yarn A is 0.61898 but for yarn B (not shown) it is 3.3903; no transformation is considered necessary.

To complete the analysis we need the sum of square between yarns. This is obtained by amending the above instructions, deleting the **transform** paragraph and changing the **histogram** paragraph. The amended instructions are:

```
/problem      title = 'example q.  cotton yarn.
              between yarns.  bmdp7d'.
/input        variables = 3.  format = free.
/variable     names = yarn, bobbin, brk_load.
/group        codes(yarn) = 1, 2.
              names(yarn) = yarna, yarnb.
/histogram    grouping = yarn.  variable =  brk_load.
/print        plot.
/end
1 1 15.0                                    48 lines of data as before
1 1 17.0

   . . .
```

Output Q.4 shows the analysis of variance. All the required sums of squares (as in Table Q.4, *App. Stat.*) are thus obtained. The output includes also a histogram for each yarn (not shown).

Alternatively, the nested analysis of variance table can be obtained using BMDP8V. The data must be rearranged, only the values of breaking load being read. As before, we first analyse each yarn separately. The instructions for yarn A are:

```
/problem     title = 'example q.  cotton yarn.
             yarn a. bmdp8v'.
/input       variables = 6.  format = free.
/design      levels = 4, 6.                        4 lengths, 6 bobbins
             names = lengths, bobbins.
             random = lengths, bobbins.            lengths nested within bobbins
             model = 'b, l(b)'.

/end
15.0 15.7 14.8 14.9 13.0 15.9                       4 lines of data for yarn A
17.0 15.6 15.8 14.2 16.2 15.6

        . . .
```

Similarly for yarn B. Output Q.5 shows the analysis of variance and estimates
of variance components for yarn A.

The nested analysis of variance for yarns, bobbins and lengths (between and
within bobbins pooled over yarns) is given by:

```
/problem     title = 'example q.  cotton yarn.
             yarn a,b. bmdp8v'.
/input       variables = 6.  format = free.
/design      levels = 2, 4, 6.                     2 yarns, 4 lengths, 6 bobbins
             names = yarn, lengths, bobbins.
             random = lengths, bobbins.
             fixed = yarn.                          bobbins nested within yarns and
             model = 'y, b(y), l(yb)'.              lengths within bobbins
/end
15.0 15.7 14.8 14.9 13.0 15.9                       8 lines of data
17.0 15.6 15.8 14.2 16.2 15.6

        . . .
```

Output Q.6 shows the analysis of variance and estimates of variance
components.

Other program

BMDP3V is a general mixed model analysis of variance program which fits
the model by maximum likelihood.

Suggested further work

Examine residuals and check on the assumption of normality. (BMDP2V
prints residuals if **print.** is added to the **design** paragraph.)

Output Q.1

Histograms, means and standard deviations. Yarn A. BMDP7D

```
                 *************                  *************
HISTOGRAM OF * brk_load *  ( 3)  GROUPED BY  * bobbin *  ( 2)
                 *************                  *************

              ONE       TWO      THREE     FOUR      FIVE      SIX
MIDPOINTS.....+.........+.........+.........+.........+.........+.........+
    18.500)                                                              *
    18.000)
    17.500)
    17.000)*
    16.500)             *                  *
    16.000)            **          *        M
    15.500)M            N          M*                           M*
    15.000)*           **          *      **       *            *
    14.500)M                                M
    14.000)*                               *         M
    13.500)                                          *
    13.000)
    12.500)

GROUP MEANS ARE DENOTED BY M'S IF THEY COINCIDE WITH *'S, N'S OTHERWISE

MEAN      15.325    16.500    16.200    14.225    15.100    15.500
STD.DEV.   1.325     1.003     1.433     1.014     1.571     0.374
```

Output Q.2

Analysis of variance: between and within bobbins. Yarn A. BMDP7D

ANALYSIS OF VARIANCE TABLE FOR MEANS					
SOURCE	SUM OF SQUARES	DF	MEAN SQUARE	F VALUE	TAIL PROBABILITY
bobbin	13.2100	5	2.6420	1.88	0.1488
ERROR	25.3550	18	1.4086		

Output Q.3

Diagnostic plots. Yarn A. BMDP7D

Output Q.4

Analysis of variance: between and within yarns. BMDP7D

ANALYSIS OF VARIANCE TABLE FOR MEANS

SOURCE	SUM OF SQUARES	DF	MEAN SQUARE	F VALUE	TAIL PROBABILITY
yarn	8.9269	1	8.9269	5.06	0.0293
ERROR	81.1613	46	1.7644		

Output Q.5

Analysis of variance: between and within bobbins. Yarn A. BMDP8V

ANALYSIS OF VARIANCE FOR DEPENDENT VARIABLE 1

SOURCE	ERROR TERM	SUM OF SQUARES	D.F.	MEAN SQUARE	F	PROB.	EXPECTED MEAN SQUARE
1 MEAN	bobbins	5747.4151	1	5747.415	2175.40	0.0000	24(1) + 4(2) + (3)
2 bobbins	I(b)	13.2100	5	2.642	1.88	0.1488	4(2) + (3)
3 I(b)		25.3550	18	1.409			(3)

ESTIMATES OF VARIANCE COMPONENTS

(1) 239.36554
(2) 0.30835
(3) 1.40861

Output Q.6

Analysis of variance: between yarns, between bobbins, within yarns and within bobbins. BMDP8V

ANALYSIS OF VARIANCE DESIGN

	INDEX		
	yarn	lengths	bobbins
NUMBER OF LEVELS	2	4	6
POPULATION SIZE	2	INF	INF

MODEL   y,b(y),l(yb)

NUMBER OF CASES READ.   8

ANALYSIS OF VARIANCE FOR DEPENDENT VARIABLE 1

SOURCE	ERROR TERM	SUM OF SQUARES	D.F.	MEAN SQUARE	F	PROB.	EXPECTED MEAN SQUARE
1 MEAN	b(y)	12144.4220	1	12144.4220	2978.12	0.0000	48(1) + 4(3) + (4)
2 yarn	b(y)	8.9269	1	8.9269	2.19	0.1698	24(2) + 4(3) + (4)
3 b(y)	l(yb)	40.7788	10	4.0779	3.64	0.0020	4(3) + (4)
4 l(yb)		40.3825	36	1.1217			(4)

ESTIMATES OF VARIANCE COMPONENTS

(1)	252.92384
(2)	0.20204
(3)	0.73903
(4)	1.12174

Example R

Biochemical experiment on the blood of mice

Description of data*

In an experiment on the effect of treatments A and B on the amount of substance S in mice's blood, it was not practicable to use more than 4 mice on any one day. The treatments formed a 2×2 system:

A_0: A absent, B_0: B absent,
A_1: A present, B_1: B present.

The mice used on one day were all of the same sex. The data are given in Table R.1 (Cox, 1958, §7.4).

Table R.1. Amount of substance S

Day 1	Male	A_0B_1	4.8	A_1B_1	6.8	A_0B_0	4.4	A_1B_0	2.8
2	Male	A_0B_0	5.3	A_1B_0	3.3	A_0B_1	1.9	A_1B_1	8.7
3	Female	A_1B_1	7.2	A_0B_1	4.3	A_0B_0	5.3	A_1B_0	7.0
4	Male	A_0B_0	1.8	A_1B_1	4.8	A_1B_0	2.6	A_0B_1	3.1
5	Female	A_1B_1	5.1	A_0B_0	3.7	A_1B_0	5.9	A_0B_1	6.2
6	Female	A_1B_0	5.4	A_0B_1	5.7	A_1B_1	6.7	A_0B_0	6.5
7	Male	A_0B_1	6.2	A_1B_1	9.3	A_0B_0	5.4	A_1B_0	6.9
8	Female	A_0B_0	5.2	A_1B_1	7.9	A_1B_0	6.8	A_0B_1	7.8

The analysis

We note that each treatment combination (A_0B_0, A_1B_0, A_0B_1 and A_1B_1) is tested on each day, but male rats are used on days 1, 2, 4 and 7 and female rats on days 3, 5, 6 and 8. Thus comparisons between treatments within males or females are independent of systematic differences between days, but not so for any comparison of treatments across the sexes. Two different error variances are involved, demanding separate estimation.

It is helpful for initial inspection to rearrange the data grouping by sex and

---

* Fictitious data based on a real investigation.

111

by treatment as in Table R.2, *App. Stat.*, reproduced here. Days are nested within sex.

Table R.2. Data rearranged according to treatments

Sex	Day	A_0B_0	A_1B_0	A_0B_1	A_1B_1
Male	1	4.4	2.8	4.8	6.8
	2	5.3	3.3	1.9	8.7
	4	1.8	2.6	3.1	4.8
	7	5.4	6.9	6.2	9.3
	Mean	4.22	3.90	4.00	7.40
Female	3	5.3	7.0	4.3	7.2
	5	3.7	5.9	6.2	5.1
	6	6.5	5.4	5.7	6.7
	8	5.2	6.8	7.9	7.9
	Mean	5.18	6.28	6.02	6.72
Overall mean		4.70	5.09	5.01	7.06

Program

BMDP8V is used. It is important to arrange the data in a systematic layout with respect to the design factors. We read the 32 observations in the order given in Table R.2, *App. Stat.*, reading across the rows. Thus data for two levels of A are read, then two levels of B, four levels of day and lastly two levels of sex. The order is specified by the **names** and **levels** statements in the **design** paragraph. The instructions for BMDP8V are as follows:

```
/problem    title = 'example r.  biochemical expt.
            bmdp8v'.
/input      variables = 4.  format = free.
/variable   names = a0b0, a1b0, a0b1, a1b1.
/design     names = sex, day, b, a.
            levels = 2, 4, 2, 2.

            model = 's, a, b, d(s)'.
            fixed = sex, day, b, a.
            print = sab.
/end
4.4 2.8 4.8 6.8
5.3 3.3 1.9 8.7
    . . .
```

2 levels of A read first, then 2 of B, . . .

sex A, B are crossed; days nested within sex
print   marginal means

8 lines of data
See: Table R.2, *App. Stat.*

Output includes all marginal means for sex $\times$ A $\times$ B; these are shown in Output R.1. The analysis of variance is shown in Output R.2. This agrees with the analysis of variance in Table R.3, *App. Stat.*, except that there the three interactions days $\times$ B (within sex), days $\times$ A (within sex) and days $\times$ B $\times$ A (within sex), each with 6 d.f., are pooled to give a residual mean square of 1.33 with 18 d.f.; this estimate of error is used for any comparisons which are independent of differences between days. Comparisons across the days will require the use of days (within sex), with 6 d.f.

Other program

BMDP2V computes analysis of variance for a factorial arrangement. The use of **exclude** in the **design** paragraph (see Example N) can specify a model having the above residual with 18 d.f. Days within sexes can be obtained by pooling days and days × sex.

Suggested further work

(i) Examine the assumption of normality. (BMDP2V will list residuals.)
(ii) See whether a transformation will reduce the sex × B × A interaction and hence simplify the conclusions.
(iii) Consider treating days as a random effect by specifying **random = day. fixed = sex, b, a.** in the **design** paragraph. The analysis of variance table (Output R.2) will then identify the different error to be used. What justification, if any, is there for treating days as random?
(iv) Estimate a standard error for the comparison of a particular treatment combination (A_1B_1, say) across the sexes.

114

Example R

Output R.1

Marginal means. BMDP8V

GRAND MEAN   5.46563

CELL AND MARGINAL MEANS

```
    s =     1         2
          4.88125   6.05000

    b =     1         2
          4.89375   6.03750

    a =     1         2
          4.85625   6.07500

    d =     1         2         3         4
s = 1    4.70000   4.80000   3.07500   6.95000
    2    5.95000   5.22500   6.07500   6.95000

    b =     1         2
s = 1    4.06250   5.70000
    2    5.72500   6.37500

    a =     1         2
s = 1    4.11250   5.65000
    2    5.60000   6.50000

    a =     1         2
b = 1    4.70000   5.08750
    2    5.01250   7.06250

    s = 1
    a =     1         2
b = 1    4.22500   3.90000
    2    4.00000   7.40000

    s = 2
    a =     1         2
b = 1    5.17500   6.27500
    2    6.02500   6.72500
```

Output R.2

Analysis of variance. BMDP8V

	SOURCE	ERROR TERM	SUM OF SQUARES	D.F.	MEAN SQUARE
	EXPECTED MEAN SQUARE				
1	MEAN		955.93782	1	955.93782
2	sex		10.92781	1	10.92781
3	b		10.46531	1	10.46531
4	a		11.88281	1	11.88281
5	d(s)		36.33188	6	6.05531
6	sb		1.95031	1	1.95031
7	sa		0.81281	1	0.81281
8	ba		5.52781	1	5.52781
9	db(s)		3.60188	6	0.60031
10	da(s)		6.19187	6	1.03198
11	sba		8.50781	1	8.50781
12	dba(s)		14.23187	6	2.37198

Example S Voltage regulator performance

Description of data

Voltage regulators fitted to private motor cars were required to operate within the range of 15.8 to 16.4 volts, and the following investigation (Desmond, 1954) was conducted to estimate the pattern of variability encountered in production. Normal procedure was for a regulator from the production line to be passed to one of a number of setting stations, where the regulator was adjusted on a test rig. These regulators then passed to one of four testing stations, where the regulator was tested, and if found to be unsatisfactory, it was passed down the production line to be reset. For the data of Table S.1, a random sample of four setting stations took part, and a number of regulators from each setting station were passed through each testing station. One special aspect of interest concerned the percentage of regulators that would be unsatisfactory were the mean kept constant at 16.1.

The analysis

Inspection across the readings for each regulator shows close consistency, except for regulator J_2 and to a lesser extent B_2. We use BMDP2V to compute a separate two-way (testing stations × regulators) analysis of variance for each of the ten setting stations A, . . . , K. The program prints residuals and these confirm the anomalous behaviour of J_2. Subsequent analysis is therefore done both with and without J_2.

The ten analyses are plotted, adding across setting stations the sums of squares and degrees of freedom for regulators (within setting stations) and for error. A further run with BMDP2V for setting stations × testing stations completes the analysis of variance (Table S.3, *App. Stat.*).

Estimates of the components of variance between setting stations and between regulators are obtained by equating observed and expected mean squares. Then assuming normality, approximate confidence limits for the percentage of regulators can be calculated as discussed in *App. Stat.*

115

Table S.1. Regulator voltages

Setting station	Regulator number	Testing station				Setting station	Regulator number	Testing station			
		1	2	3	4			1	2	3	4
A	1	16.5	16.5	16.6	16.6	F	1	16.1	16.0	16.0	16.2
	2	15.8	16.7	16.2	16.3		2	16.5	16.1	16.5	16.7
	3	16.2	16.5	15.8	16.1		3	16.2	17.0	16.4	16.7
	4	16.3	16.5	16.3	16.6		4	15.8	16.1	16.2	16.2
	5	16.2	16.1	16.3	16.5		5	16.2	16.1	16.4	16.2
	6	16.9	17.0	17.0	17.0		6	16.0	16.2	16.2	16.1
	7	16.0	16.2	16.0	16.0		11	16.0	16.0	16.1	16.0
	11	16.0	16.0	16.1	16.0	G	1	15.5	15.5	15.3	15.6
							2	16.0	15.6	15.7	16.2
B	1	16.0	16.1	16.0	16.1		3	16.0	16.4	16.2	16.2
	2	15.4	16.4	16.8	16.7		4	15.8	16.5	16.2	16.2
	3	16.1	16.4	16.3	16.3		5	15.9	16.1	15.9	16.0
	4	15.9	16.1	16.0	16.0		6	15.9	16.1	15.8	15.7
							7	16.0	16.4	16.0	16.0
C	1	16.0	16.0	15.9	16.3		12	16.1	16.2	16.2	16.1
	2	15.8	16.0	16.3	16.0						
	3	15.7	16.2	15.3	15.8	H	1	15.5	15.6	15.4	15.8
	4	16.2	16.4	16.4	16.6		2	15.8	16.2	16.0	16.2
	5	16.0	16.1	16.0	15.9		3	16.2	15.4	16.1	16.3
	6	16.1	16.1	16.1	16.1		4	16.1	16.2	16.0	16.1
	10	16.1	16.0	16.1	16.0		5	16.1	16.2	16.3	16.2
							10	16.1	16.1	16.0	16.1
D	1	16.1	16.0	16.0	16.1						
	2	16.0	15.9	16.2	16.0	J	1	16.2	16.1	15.8	16.0
	3	15.7	15.8	15.7	15.7		2	16.2	15.3	17.8	16.3
	4	15.6	16.4	16.1	16.2		3	16.4	16.7	16.5	16.5
	5	16.0	16.2	16.1	16.1		4	16.2	16.5	16.1	16.1
	6	15.7	15.7	15.7	15.7		5	16.1	16.4	16.1	16.3
	11	16.1	16.1	16.1	16.0		10	16.4	16.3	16.4	16.4
						K	1	15.9	16.0	15.8	16.1
E	1	15.9	16.0	16.0	16.5		2	15.8	15.7	16.7	16.0
	2	16.1	16.3	16.0	16.0		3	16.2	16.2	16.2	16.3
	3	16.0	16.2	16.0	16.1		4	16.2	16.3	15.9	16.3
	4	16.3	16.5	16.4	16.4		5	16.0	16.0	16.0	16.0
							6	16.0	16.4	16.2	16.2
							11	16.0	16.1	16.0	16.1

Program

BMDP2V is used to compute the analyses of variance and to list residuals. Since the variation in the data is small relative to the observed mean voltage, minor, although nontrivial, numerical rounding errors arise in the computations if the raw data are used. We work therefore with voltage minus 16. This is calculated in the **transform** paragraph. Separate two-way analyses are required initially for each setting station. Instructions for setting station 1 are given below. For setting stations 2, . . . , 10 amend the sentence **use = sstation eq 1.** accordingly.

```
/problem      title = 'example s.  voltage regulators.
              separate setting stations.  bmdp2v.'.
/input        variables = 4.  format = free.
/variable     names = sstation, regulat, tstation, y.
/transform    use = sstation eq 1.
              y = y - 16.                              voltage minus 16
/design       dependent = y.  grouping = regulat,      two-way regs. × tstations
              tstation.
              print.                                   print residuals
/end
1 1 1 16.5                                             256 lines of data
1 1 2 16.5                                             Code:    A = 1,    B = 2, . . . ,
   . . .                                                        K = 10 (col. 1)
```

Note: the **for–%** option may be used to make the program repeat automatically for all ten setting stations but to do so requires the data to be on a separate file; see note 9(ii) in the introductory section (p. 9).

The output gives an analysis of variance table and a list of residuals for each of the ten setting stations. Details for setting station J are shown in Output S.1(a) and (b). The other setting stations (not shown here) have smaller error mean square. The residuals in Output S.1(b) confirm the inconsistency in J_2.

To repeat the above analysis for setting station J omitting J_2, we amend the instructions, changing the **use** statement of the **transform** paragraph, as follows:

```
/problem      title = 'example s.  voltage regulators.
              sstation j, omitting j2.  bmdp2v'.
/input        variables = 4.  format = free.
/variable     names = sstation, regulat, tstation, y.
/transform    use = sstation eq 9 and regulat ne 2.
              y = y - 16.
/design       dependent = y.
              grouping = regulat, tstation.
/end
1 1 1 16.5                                    256 lines of data as before
1 1 2 16.5

   . . .
```

The analysis of variance is shown in Output S.2; the error mean square is reduced to 0.0160, comparable to the variation in the other nine setting stations (Table S.2, *App. Stat.*).

Instructions for computing the analysis of variance for setting stations × testing stations are as follows:

```
/problem      title = 'example s.  voltage regulators.
              sstation x tstation.  bmdp2v'.
/input        variables = 4.  format = free. ;
/variable     names = sstation, regulat, tstation, y.
/transform    y = y - 16.
/design       dependent = y.
              grouping = sstation, tstation.
/end
1 1 1 16.5                                    256 lines of data as before
1 1 2 16.5

   . . .
```

Output S.3 shows the analysis of variance.

For the corresponding analysis of variance omitting regulator J_2, we add a sentence to the **transform** paragraph as follows:

```
/problem     title = 'example s.  voltage regulators.
             sstation x tstation, omitting j2.
             bmdp2v'.
/input       variables = 4.  format = free.
/variable    names = sstation, regulat, tstation, y.
/transform   y = y - 16.
             if (sstation eq 9 and regulat eq 2)      omit regulator J₂
             then use = 0.
/design      dependent = y.
             grouping = sstation, tstation.
/end
1 1 1 16.5                                            256 lines of data as before
1 1 2 16.5
    ...
```

Output S.4 shows the analysis of variance with J_2 omitted. (In early printings Table S.3, *App. Stat.* contains an error in the testing station s.s.)

The error sum of squares with 216 d.f. in Output S.3 is equal to the total of the two components:

(i) regulators within setting stations, with 54 d.f.;
(ii) residual, with 162 d.f.;

where the sums of squares for (i) and (ii) are obtained by pooling those from the ten analyses within setting stations (as in Output S.1). That with 212 d.f. (in Output S.4) is the sum of corresponding terms when J_2 is omitted.

Suggested further work

(i) Examine the assumption of normality.
(ii) Check for any other outliers.

Output S.1

Setting station J, full data. BMDP2V
(a) Analysis of variance

SOURCE	SUM OF SQUARES	DEGREES OF FREEDOM	MEAN SQUARE	F	TAIL PROB.
MEAN	2.10042	1	2.10042	9.43	0.0078
regulat	0.61208	5	0.12242	0.55	0.7363
tstation	0.19792	3	0.06597	0.30	0.8275
ERROR	3.33958	15	0.22264		

(b) Predicted values and residuals

CASE	regulat	tstation	PREDICTD	RESIDUAL
1	*1	*1	-0.02083	0.22083
2	*1	*2	-0.05417	0.15417
3	*1	*3	0.17917	-0.37917
4	*1	*4	-0.00417	0.00417
5	*2	*1	0.35417	-0.15417
6	*2	*2	0.32083	-1.02083
7	*2	*3	0.55417	1.24583
8	*2	*4	0.37083	-0.07083
9	*3	*1	0.47917	-0.07917
10	*3	*2	0.44583	0.25417
11	*3	*3	0.67917	-0.17917
12	*3	*4	0.49583	0.00417
13	*4	*1	0.17917	0.02083
14	*4	*2	0.14583	0.35417
15	*4	*3	0.37917	-0.27917
16	*4	*4	0.19583	-0.09583
17	*5	*1	0.17917	-0.07917
18	*5	*2	0.14583	0.25417
19	*5	*3	0.37917	-0.27917
20	*5	*4	0.19583	0.10417
21	*10	*1	0.32917	0.07083
22	*10	*2	0.29583	0.00417
23	*10	*3	0.52917	-0.12917
24	*10	*4	0.34583	0.05417

Similar tables are output for other setting stations.

Output S.2

Setting station J, J_2 omitted. Analysis of variance. BMDP2V

SOURCE	SUM OF SQUARES	DEGREES OF FREEDOM	MEAN SQUARE	F	TAIL PROB.
MEAN	1.51250	1	1.51250	94.53	0.0000
regulat	0.56000	4	0.14000	8.75	0.0015
tstation	0.12550	3	0.04183	2.61	0.0993
ERROR	0.19200	12	0.01600		

120

Example S

Output S.3

Setting station × testing station, full data. Analysis of variance. BMDP2V

SOURCE	SUM OF SQUARES	DEGREES OF FREEDOM	MEAN SQUARE	F	TAIL PROB.
MEAN	3.92866	1	3.92866	46.52	0.0000
sstation	4.41905	9	0.49101	5.81	0.0000
tstation	0.86067	3	0.28689	3.40	0.0187
st	0.90406	27	0.03348	0.40	0.9971
ERROR	18.24220	216	0.08445		

Output S.4

Setting station × testing station, J_2 omitted. Analysis of variance. BMDP2V

SOURCE	SUM OF SQUARES	DEGREES OF FREEDOM	MEAN SQUARE	F	TAIL PROB.
MEAN	3.72624	1	3.72624	52.52	0.0000
sstation	4.16209	9	0.46245	6.52	0.0000
tstation	0.94241	3	0.31414	4.43	0.0048
st	0.72077	27	0.02670	0.38	0.9982
ERROR	15.04254	212	0.07096		

Example T   Intervals between the failure of air-conditioning equipment in aircraft

Description of data

The data in Table T.1, reported by Proschan (1963), are the intervals in service-hours between failures of the air-conditioning equipment in 10 Boeing 720 jet aircraft. It is required to describe concisely the variation within and between aircraft, with emphasis on the forms of the frequency distributions involved.

The analysis

The data can be analysed in various ways to check consistency with an assumed theoretical model and to make comparisons between aircraft. Here we concentrate upon fitting by maximum likelihood a gamma distribution to the observed time intervals for each aircraft.

The gamma distribution of mean μ and index β,

$$\frac{(\beta/\mu)\,(\beta y/\mu)^{\beta-1}e^{-\beta y/\mu}}{\Gamma(\beta)} \qquad \text{(T.2, App. Stat.)}$$

reduces to the exponential distribution when $\beta = 1$, and comparisons between aircraft then become simply comparisons between means. Hence it is sensible not only to check on the consistency of the estimated β's but also to compare their common estimate with $\beta = 1$. Instructions are given for fitting:

(i) separate gamma distributions to all aircraft, i.e. parameters μ_i, β_i for $i = 1, \ldots, 10$;

(ii) separate gamma distributions but with common β.

The maximum likelihood estimate of μ_i is given by $\hat{\mu}_i = \bar{y}_i$, the sample mean, but it is necessary to iterate to obtain the estimates $\hat{\beta}_i$ or $\hat{\beta}$. It would be possible, but less efficient, to iterate simultaneously for $\hat{\mu}_i$ and $\hat{\beta}_i$. We choose here to iterate only for the $\hat{\beta}_i$'s (and $\hat{\beta}$).

121

Table T.1. Intervals between failures (operating hours)

Aircraft number

1	2	3	4	5	6	7	8	9	10
413	90	74	55	23	97	50	359	487	102
14	10	57	320	261	51	44	9	18	209
58	60	48	56	87	11	102	12	100	14
37	186	29	104	7	4	72	270	7	57
100	61	502	220	120	141	22	603	98	54
65	49	12	239	14	18	39	3	5	32
9	14	70	47	62	142	3	104	85	67
169	24	21	246	47	68	15	2	91	59
447	56	29	176	225	77	197	438	43	134
184	20	386	182	71	80	188		230	152
36	79	59	33	246	1	79		3	27
201	84	27	15	21	16	88		130	14
118	44	153	104	42	106	46			230
34	59	26	35	20	206	5			66
31	29	326		5	82	5			61
18	118			12	54	36			34
18	25			120	31	22			
67	156			11	216	139			
57	310			3	46	210			
62	76			14	111	97			
7	26			71	39	30			
22	44			11	63	23			
34	23			14	18	13			
	62			11	191	14			
	130			16	18				
	208			90	163				
	70			1	24				
	101			16					
	208			52					
				95					

As an approximation for $\Gamma(\beta)$ when $\beta > 1$, we use

$$e^{-\beta}\beta^{\beta-\frac{1}{2}}(2\pi)^{\frac{1}{2}}\left\{1 + \frac{1}{12\beta} + \frac{1}{288\beta^2} - \frac{139}{51840\beta^3} \cdots\right\}$$

which together with $\Gamma(\beta + 1) = \beta\Gamma(\beta)$ gives an approximation in our required range $0.4 < \beta < 1.8$. Within this range it is accurate to 0.0001.

Program

BMDPAR, a nonlinear least squares program, is used. If **xloss**, equal to minus the log likelihood, is specified in the **fun** paragraph and **loss** written into the regress paragraph, the program will calculate maximum likelihood estimates.

We need to fit a separate gamma distribution for each aircraft to estimate $\beta_1, \ldots, \beta_{10}$. Instructions for aircraft 1 are given below. For aircraft 2, . . . , 10 amend the two sentences **use = aircraft eq 1.** and **mu = 95.7.** appropriately; the values of **mu** for aircraft 2, . . . , 10 are 83.5, 121.3, 130.0, 59.6, 76.8, 64.1, 200.0, 108.1, 82.0, respectively.

```
/problem     title = 'example t.  air-conditioning.
             individual beta.  aircraft 1.  bmdpar'.
/input       variables = 2.  format = free.
/variable    names = aircraft, time.  add = new.
/transform   dummy = 1.0.
             use = aircraft eq 1.                      aircraft 1 only
/regress     dependent = dummy.
             parameters = 1.  print = 0.  loss.
/parameter   initial = 1.0.                            initial approx. for β
/fun         u = p1 + 1.0.                             p1 = β
             c1 = 1.0/12.0.
             c2 = 1.0/288.0.
             c3 = -139.0/51840.0.
             g = -ln(p1) - u + (u - 0.5)*ln(u) +
             mu = 95.7.                                g = log Γ(β) − log(2π)^½
             ln(((c3/u + c2)/u + c1)/u + 1.0).         μ₁ = 95.7
             f = p1*ln(p1) - p1*ln(mu) - g +
             (p1 - 1.0)*ln(time) - p1*time/mu.
             xloss = -f.                               xloss = −log likelihood + n
                                                          log(2π)^½
             dummy = f + 1.0.                          dummy recalculated
/end
1 413                                                  199 lines of data.
1 14                                                   col. 1 = aircraft no.
...
```

In the right-margin annotations:

$\mathbf{p1} = \beta$

$\mathbf{g} = \log \Gamma(\beta) - \log(2\pi)^{\frac{1}{2}}$
$\mu_1 = 95.7$

$\mathbf{xloss} = -\log \text{ likelihood } + n$
$\log(2\pi)^{\frac{1}{2}}$

Note: the **for–%** option may be used to make the program repeat automatically for all ten aircraft, but to do so requires the data to be on a separate file and its name stated in the **input** paragraph. After the **problem** paragraph insert

for k = 1 to 10.
    mu = 95.7,83.5,121.3,130.0,59.6,76.8,64.1,200.0,108.1,82.0.
%

and amend the sentence in the **transform** paragraph to

use = aircraft eq k.

See note 9(ii) in the introductory section (p. 9).

The value of the loss function, i.e. $-\log$ likelihood $+ n \log(2\pi)^{\frac{1}{2}}$, and the estimate $\hat{\beta}_1$ for aircraft 1 are shown in Output T.1. Note that the mean time interval is also printed, giving a check on the value read in above for $\hat{\mu}_1$. (The estimate $\hat{\beta}$ for aircraft 3 is incorrect in early printings of *App. Stat.*)

To fit gamma distributions with a common index β the instructions are amended as follows:

```
/problem     title = 'example t.  air-conditioning.
             common beta.  bmdpar'.
/input       variables = 2.  format = free.
/variable    names = aircraft, time.  add = new.
/transform   dummy = 1.0.
/regress     dependent = dummy.
             parameters = 1.  print = 0.  loss.
/parameter   initial = 1.0.
/fun         u = p1 + 1.0.
             c1 = 1.0/12.0.
             c2 = 1.0/288.0.
             c3 = -139.0/51840.0.
             g = -ln(p1) - u + (u - 0.5)*ln(u)+
             ln(((c3/u + c2)/u + c1)/u + 1.0).
             mu = rec(aircraft, 1, 95.7,             recode mu for aircraft 1, ..., 10
             2, 83.5, 3, 121.3, 4, 130.9, 5, 59.6,
             6, 76.8, 7, 64.1, 8, 200.0,
             9, 108.1, 10, 82.0).
             f = p1*ln(p1) - p1*ln(mu) - g +
             (p1 - 1.0)*ln(time) - p1*time/mu.
             meansquare = 1.0.
             xloss = -f.
             dummy = f + 1.0.
/end
1 413                                                  199 lines of data as before
1 14

...
```

Output T.2 shows the loss function and estimate $\hat{\beta}$. The hypothesis of a common index β can be tested by differencing the total of the loss functions for separate β_i's and the loss function for common β, and comparing the difference with $\frac{1}{2}\chi^2$ with 9 degrees of freedom.

Aircraft 8 will be found to have a low value of $\hat{\beta}$ ($=0.46$). To exclude this aircraft from the calculations when estimating a common β, write

use = aircraft ne 8.

into the **transform** paragraph. This yields an estimate $\hat{\beta} = 1.07$.

For an interpretation of the conclusions from the above analysis, see the discussion in *App. Stat.*; the conclusions are not clear-cut.

Other program

BMDP3R is also a nonlinear iterative least squares program which will give maximum likelihood estimates. It requires the derivatives of the likelihood and this will necessitate a series approximation for the derivative of log $\Gamma(\beta)$, as well as for $\Gamma(\beta)$.

Suggested further work

(i) Rerun the analysis iterating on $(\hat{\mu}, \hat{\beta})$ instead of just $\hat{\beta}$. How much slower is the iteration?

(ii) If the gamma distribution is reparameterized as $\rho^\beta y^{\beta-1} e^{-\rho y}/\Gamma(\beta)$ with iteration on $(\hat{\rho}, \hat{\beta})$, how is the speed of iteration affected?

(iii) Carry out a similar analysis using the Weibull distribution and compare your results.

Output T.1

Loss function and maximum likelihood estimate $\hat{\beta}_1$ for aircraft 1. BMDPAR

VARIABLE NO. NAME		MEAN	STANDARD DEVIATION	MINIMUM	MAXIMUM
1	aircraft	1.000000	0.000000	1.000000	1.000000
2	time	95.695656	119.289726	7.000000	447.000000
3	dummy	1.000000	0.000000	1.000000	1.000000

ITER. NO.	INCR. HALV.	LOSS FUNCTION	PARAMETERS P1
0	0	106.894068	1.100000
0	0	106.771223	1.000000
1	3	106.764172	0.982649
2	1	106.762016	0.959537
3	2	106.761832	0.962191
4	1	106.761767	0.966151
5	2	106.761762	0.965687
6	1	106.761760	0.965018
7	2	106.761760	0.965095

THE LOSS FUNCTION (=   106.762   ) WAS SMALLEST WITH THE FOLLOWING PARAMETER VALUES

PARAMETER	ESTIMATE	ASYMPTOTIC STANDARD DEVIATION	COEFFICIENT OF VARIATION
P1	0.965095	0.743254	0.770136

Similar tables are output for aircraft 2 to 10.

Output T.2

Loss function and maximum likelihood estimate $\hat{\beta}$ (common to all aircraft). BMDPAR

ITER. NO.	INCR. HALV.	LOSS FUNCTION	PARAMETERS P1
0	0	904.183083	1.100000
0	0	903.665293	1.000000
1	4	903.663284	1.003410
2	1	903.662744	1.006518
3	4	903.662742	1.006403
4	1	903.662742	1.006319
5	2	903.662742	1.006328

THE LOSS FUNCTION (=   903.663   ) WAS SMALLEST WITH THE FOLLOWING PARAMETER VALUES

PARAMETER	ESTIMATE	ASYMPTOTIC STANDARD DEVIATION	COEFFICIENT OF VARIATION
P1	1.006328	0.219462	0.218082

Example U

Survival times of leukemia patients

Description of data

The data in Table U.1 from Feigl and Zelen (1965) are time to death, Y, in weeks from diagnosis and $\log_{10}$ (initial white blood cell count), x, for 17 patients suffering from leukemia. The relation between Y and x is the main aspect of interest.

Table U.1. Survival time Y in weeks and $\log_{10}$ (initial white blood cell count), x, for 17 leukemia patients

x	Y	x	Y	x	Y
3.36	65	4.00	121	4.54	22
2.88	156	4.23	4	5.00	1
3.63	100	3.73	39	5.00	1
3.41	134	3.85	143	4.72	5
3.78	16	3.97	56	5.00	65
4.02	108	4.51	26		

The analysis

It is sensible to choose a model for Y which is positive for all values of x. Hence we consider

$$Y_i = \beta_0 \exp\{\beta_1(x_i - \bar{x})\}\varepsilon_i \qquad \text{(U.1, App. Stat.)}$$

where ε_i is a random term assumed provisionally to be exponentially distributed with unit mean. Then $E(Y_i) = s.d.(Y_i) = \beta_0 \exp\{\beta_1(x_i - \bar{x})\}$. We fit the model by BMDP3R using the method of iterated weighted least squares, with weights equal to $\{\text{var}(Y_i)\}^{-1}$.

For BMDP3R we need the partial derivatives of $E(Y_i)$ with respect to β_0 and β_1, which are respectively $\exp\{\beta_1(x_i - \bar{x})\}$ and $\beta_0(x_i - \bar{x})\exp\{\beta_1(x_i - \bar{x})\}$.

Taking the model in the above form leads to zero covariance between the estimates $\hat{\beta}_0$ and $\hat{\beta}_1$ which is helpful if determining a standard error for

127

predicted values of *Y* and which with more complex sets of data would have numerical analytical advantages.

Program

BMDP3R is used to fit the model by the method of iterated weighted least squares. In the **fun** paragraph it is necessary to write

(i) $f = E(Y_i)$.
(ii) **df1** and **df2** for the derivatives with respect to β_0, β_1.
(iii) $w = 1/f^2$ for the weight.
(vi) $p1 = \beta_0$ and $p2 = \beta_1$ denoting the unknown parameters.

See Example H, or BMDP manual (Dixon *et al.*, 1985), for comments on the **regress** paragraph. The instructions for using BMDP3R to fit the above model are as follows:

```
/problem     title = 'example u.
             leukemia patients.  bmdp3r'.
/input       variables = 2.  format = free.
/variable    names = lwbc, weeks.  add = new.
/transform   w = 1.0.
/regress     dependent = weeks.
             parameters = 2.  weight = w.
             halving = 0.  meansquare = 1.0.
/parameter   initial = 40.0, -1.0.
/fun         u = exp(p2*(lwbc - 4.0959)).           x̄ = 4.0959
             f = p1*u.                              f = E(Y)
             df1 = u.                               df1 = ∂F/∂β₀
             df2 = f*(lwbc - 4.0959).               df2 = ∂F/∂β₁
             w = 1/(f**2).                          w = weight
/plot        variable = lwbc.                       plot fitted model
/end
3.36 65                                             17 lines of data
2.88 156
. . .
```

The default value for initial values for β_0 and β_1 is zero; instead we specify them as 40.0 and -1.0.

Output U.1 shows the estimates $\hat{\beta}_0$ and $\hat{\beta}_1$ and their asymptotic standard errors. They agree with those fitted by maximum likelihood in *App. Stat.* Output U.2 lists the predicted and observed values and shows them plotted against *x* (= **lwbc**). The residuals listed in Output U.2 are irrelevant, it being better to define residuals in this case as the ratio of the observed to fitted survival time, which can then be examined for consistency with an exponential distribution by plotting against expected exponential order statistics.

Other program

BMDPAR will fit by iterated weighted least squares and does not require derivatives to be specified.

Suggested further work

(i) Fit the model in the nonorthogonal form, i.e. $E(Y_i) = \gamma_0 \exp(\gamma_1 x_i)$, and compare with preceding analysis.

(ii) Determine a confidence interval for the expected value of Y for a given value x.

Output U.1

Estimates and standard errors. BMDP3R

ITERATION NUMBER	INCREMENT HALVINGS	RESIDUAL SUM OF SQUARES	P1	P2
0	0	22.3032	40.000000	-1.000000
1	0	14.3075	51.211838	-1.129909
2	0	14.0994	51.110878	-1.110681
3	0	14.0842	51.107160	-1.109389
4	0	14.0830	51.107143	-1.109304
5	0	14.0830	51.107143	-1.109298
6	0	14.0830	51.107143	-1.109298
7	0	14.0830	51.107143	-1.109298
8	0	14.0830	51.107143	-1.109298
9	0	14.0830	51.107143	-1.109298

ASYMPTOTIC CORRELATION MATRIX OF THE PARAMETERS

		P1	P2
		1	2
P1	1	1.0000	
P2	2	0.0000	1.0000

PARAMETER	ESTIMATE	ASYMPTOTIC STANDARD DEVIATION	TOLERANCE
P1	51.107143	12.395303	0.9999999991
P2	-1.109298	0.399654	0.9999999991

Output U.2

Predicted and observed values. BMDP3R

CASE NO. LABEL	PREDICTED weeks	STD DEV OF PRED VALUE	OBSERVED weeks	RESIDUAL
1	115.613968	44.072674	65.000000	-50.613968
2	196.904648	106.938293	156.000000	-40.904640
3	85.690857	26.201086	100.000000	14.309145
4	109.376022	40.032681	134.000000	24.623978
5	72.555489	19.838457	16.000000	-56.555485
6	55.596489	13.589133	108.000000	52.403511
7	56.843735	13.957647	121.000000	64.156265
8	44.043045	10.939761	4.000000	-40.043045
9	76.693459	21.720058	39.000000	-37.693459
10	67.134682	17.568281	143.000000	75.865318
11	58.767262	14.556565	56.000000	-2.767261
12	32.283745	9.479285	26.000000	-6.283743
13	31.227064	9.385149	22.000000	-9.227064
14	18.746519	8.158210	1.000000	-17.746519
15	18.746519	8.158210	1.000000	-17.746519
16	25.574909	8.897698	5.000000	-20.574909
17	18.746519	8.158210	65.000000	46.253483

Output U.2 (continued)

SERIAL CORRELATION   -0.134

PLOTS OF VARIABLE(1) VERSUS PREDICTED AND OBSERVED VARIABLE(2) AND VERSUS
RESIDUALS.

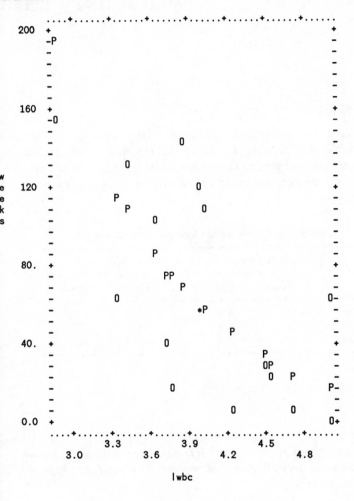

Example V   A retrospective study with binary data

Description of data

In a retrospective study of the possible effect of blood group on the incidence of peptic ulcers, Woolf (1955) obtained data from three cities. Table V.1 gives for each city data for blood groups O and A only. In each city, blood group is recorded for peptic-ulcer subjects and for a control series of individuals not having peptic ulcer.

Table V.1. Blood groups for peptic ulcer and control subjects

	Peptic ulcer		Control	
	Group O	*Group A*	*Group O*	*Group A*
London	911	579	4578	4219
Manchester	361	246	4532	3775
Newcastle	396	219	6598	5261

The analysis

As discussed in *App. Stat.*, we wish to study how the probability of a peptic ulcer depends on blood group, but the data have been collected in an inverse way from samples of peptic ulcer and non-peptic ulcer subjects. However, the log of the odds ratio is given by

$$\Delta = \log\left\{\frac{\text{pr (ulcer}|\text{A})}{\text{pr (ulcer}|\text{O})} \times \frac{\text{pr (no ulcer}|\text{O})}{\text{pr (no ulcer}|\text{A})}\right\} \qquad (\text{V.1, } App. Stat.)$$

$$= \log\left\{\frac{\text{pr (A}|\text{ulcer})}{\text{pr (O}|\text{ulcer})}\right\} - \log\left\{\frac{\text{pr (A}|\text{no ulcer})}{\text{pr (O}|\text{no ulcer})}\right\}$$

and we can estimate Δ by fitting the logistic model (taking the approximately orthogonal form with $\pm\beta_1$):

$$\log\left\{\frac{\text{pr(A)}}{\text{pr(O)}}\right\} = \begin{cases} \beta_0 + \beta_1 & \text{ulcer group} \\ \beta_0 - \beta_1 & \text{non-ulcer group} \end{cases}$$

where $\Delta = 2\beta_1$. We fit the model to each of the cities separately, to check the consistency of Δ, and then an overall model to obtain a common estimate of Δ.

The model fitted here is very simple, just a comparison of two groups, but the principle illustrated is much more general and widely used in case–control studies.

Program

BMDPLR is used to fit a logistic model by maximum likelihood to the proportion of blood group A subjects in the ulcer and no-ulcer samples. This is done separately for each city. Instructions for city 1 (London) are given below. For city 2 (Manchester) or city 3 (Newcastle) amend the sentence **use = city eq 1.** appropriately.

```
/problem     title = 'example v.  peptic ulcers.
             london.  bmdplr'.
/input       variables = 4.  format = free.
/variables   names = city, ulcer, o, a.
/transform   use = city eq 1.                    use only data for city 1.
/regress     scount = a.
             fcount = o.
             model = ulcer.
/end
1 1 911 579                                       6 lines of data.
1 0 4578 4219
                                                  Code:  col 1.  London = 1,  Man-
                                                         chester = 2, Newcastle = 3
                                                         col 2. ulcer = 1, control = 0.

   . . .
```

Note: the **for–%** option may be used to make the program repeat automatically for all three cities but to do so requires the data to be on a separate file; see note 9(ii) in the introductory section (p. 9).

The program generates design variables ± 1 to represent the ulcer and control group; see Output V.1. The estimates and standard errors for the parameters are shown in Output V.2. The estimate of Δ for London is $2 \times (-0.18579) = -0.3716$. The estimates and standard errors agree with those of Table V.2, *App. Stat.*, obtained by other large-sample approximate methods.

To obtain an estimate of Δ pooled over the three cities, we remove the **transform** paragraph and specify a model to include cities. The instructions are:

```
/problem     title = 'example v.  peptic ulcers.
             bmdplr.'.
/input       variables = 4.  format = free.
/variables   names = city, ulcer, o, a.
/regress     scount = a.
             fcount = o.
             model = ulcer, city.                    model with ulcer and city parameters
/end
1 1 911 579                                           6 lines of data as before
1 0 4578 4219

    . . .
```

Estimates for the fitted model are shown in Output V.3. Also shown is the value of $-2 \log (\text{likelihood}) = 2.965$ with 2 d.f., agreeing closely with the χ^2 statistic (V.6) in *App. Stat.* The estimated value of Δ is $2 \times (-0.16511) = -0.3302$, with standard error 0.0417.

Output V.1

Design variables for control and ulcer groups. BMDPLR

```
VARIABLE    VALUE OR        DESIGN VARIABLES
NO. N A M E  INTERVAL  FREQ    ( 1)

  2 ulcer        0     8797      -1
                 1     1490       1
```

Output V.2

Estimates and standard errors. Separate cities. BMDPLR
(a) London

```
                                STANDARD
    TERM        COEFFICIENT       ERROR     COEFF/S.E.

ulcer           -0.18579       0.2864E-01    -6.488
CONSTANT        -0.26745       0.2864E-01    -9.339
```

(b) Manchester

```
                                STANDARD
    TERM        COEFFICIENT       ERROR     COEFF/S.E.

ulcer           -0.10039       0.4278E-01    -2.347
CONSTANT        -0.28315       0.4278E-01    -6.619
```

(c) Newcastle

```
                                STANDARD
    TERM        COEFFICIENT       ERROR     COEFF/S.E.

ulcer           -0.18295       0.4311E-01    -4.244
CONSTANT        -0.40939       0.4311E-01    -9.497
```

Output V.3

Estimates and standard errors. Model over cities. BMDPLR

```
GOODNESS OF FIT CHI-SQ  (2*0*LN(O/E)) =   2.965  D.F.=  2  P-VALUE= 0.227
GOODNESS OF FIT CHI-SQ  ( D. HOSMER ) =   0.229  D.F.=  2  P-VALUE= 0.892
GOODNESS OF FIT CHI-SQ  ( C.C.BROWN ) =   0.000  D.F.=  0  P-VALUE= 1.000

                                STANDARD
    TERM        COEFFICIENT       ERROR     COEFF/S.E.

ulcer                -0.16511       0.2084E-01   -7.924
city          (1)    -0.11014E-01   0.1679E-01   -0.6558
              (2)    -0.64826E-01   0.1553E-01   -4.174
CONSTANT             -0.32837       0.2077E-01  -15.81
```

Example W

Housing and associated factors

Description of data

The data in Table W.1 (Madsen, 1976) relate to an investigation into satisfaction with housing conditions in Copenhagen. A total of 1681 residents from selected areas living in rented homes built between 1960 and 1968 were questioned on their satisfaction, the degree of contact with other residents and their feeling of influence on apartment management. The purpose of the investigation was to study association between these three factors and the type of housing.

Table W.1. 1681 persons classified according to satisfaction, contact, influence and type of housing

Contact		Low			High		
Satisfaction		Low	Medium	High	Low	Medium	High
Housing	*Influence*						
Tower blocks	Low	21	21	28	14	19	37
	Medium	34	22	36	17	23	40
	High	10	11	36	3	5	23
Apartments	Low	61	23	17	78	46	43
	Medium	43	35	40	48	45	86
	High	26	18	54	15	25	62
Atrium houses	Low	13	9	10	20	23	20
	Medium	8	8	12	10	22	24
	High	6	7	9	7	10	21
Terraced houses	Low	18	6	7	57	23	13
	Medium	15	13	13	31	21	13
	High	7	5	11	5	6	13

The analysis

Type of housing is treated as an explanatory variable, with satisfaction, contact and influence being treated as response variables.

136

We first inspect the distribution for each response variable within each type of housing. There are marked differences in distribution between types of housing. Hence we fit loglinear models in the three response variables, separately within each type of housing, but hoping to obtain reasonably consistent conclusions from the separate models.

Program

BMDP4F is used for all the analysis. The instructions are given below. The data are read in exactly the layout of Table W.1 (this is an option available for BMDP4F).

```
/problem     title = 'example w. housing. bmdp4f'.
/input       variables = 4.  table = 3, 2, 3, 4.    describes layout of data
             format = free.
/variable    names = satisfy, contact, influenc,
             housing.
/category    codes(satisfy,influenc) = 1 to 3.
             names(satisfy,influenc) = low,
             medium, high.
             codes(contact) = 1, 2.                 labels for output
             names(contact) = low, high.
             codes(housing) = 1 to 4.
             names(housing ) = tower, apt,
             atrium, terrace.
/table       indices = satisfy, contact,            print input data
             influence, housing.
/table       row = housing.
             col = satisfy, contact, influenc.      print 2-way tables H × S, H × C,
             cross.                                  H × I.
/print       percent = row.                         percentages across rows of 2-way
                 .                                   tables
/table       indices = satisfy, contact, influenc.  S,C,I = response variables
             condition = housing.                   H = explanatory variable
/fit         all.                                   hierarchy  of  models  within
                                                        S × C × I
             model = si, c.                          main effects plus S × I
             model = s, c, i.                        independence model
/print       expected.  standardized.               print fitted frequencies and resi-
                                                        duals
             percent = no.                          percentages not required
/end
21 21 28 14 19 37                                    12 lines of data
34 22 36 17 23 40
   . . .
```

The first **table** paragraph prints simply the full array of observed frequencies. The second **table** paragraph prints two-way tables for satisfaction × housing, contact × housing, and influence × housing; percentages across rows are also printed (corresponding to Table W.2, *App. Stat.*). Output W.1 gives details for satisfaction × housing.

The third **table** paragraph prints, separately within each type of housing, the three-way table S × C × I of observed frequencies. The **fit** paragraph fits models to these S × C × I tables. Output W.2 shows, for tower blocks, goodness-of-fit statistics for all models within the hierarchy containing all two-factor interactions (i.e. given by **fit all.**). Output W.3 gives, also for tower blocks, the fitted frequencies and standardized residuals for the model SI,C, i.e. containing main effects and S × I. Note that this model will not be found to fit so well for apartments; see discussion in *App. Stat.*

Fitted frequencies based on the independence model (i.e. main effects only) are used in *App. Stat.* to aid interpretation of the interaction S × I. These are given by the sentence **model = s,c,i.** in the **fit** paragraph, but the output is not shown here.

Suggested further work

(i) Insert **lambda.** into the final **print** paragraph to obtain estimates and standard errors of parameters in the loglinear model SI,C. Do these aid interpretation of data?

(ii) Try adding **cell = standardized.** and **step = 5.** to the final **fit** paragraph. This will permit the program to identify up to 5 extreme residuals and refit the model omitting the appropriate cells.

(iii) Consider ways in which the order of the categories low, medium and high for satisfaction and influence may be taken into account.

Output W.1

Distribution of respondents according to satisfaction for each type of housing. BMDP4F

```
*************************
* TABLE PARAGRAPH    2  *
*************************

*****  OBSERVED FREQUENCY TABLE  2

housing                satisfy
------                 ------
           low    medium    high    TOTAL
---------------------------------------------
tower       99      101     200 |     400
apt        271      192     302 |     765
atrium      64       79      96 |     239
terrace    133       74      70 |     277
--------------------------------|---------
TOTAL      567      446     668 |    1681

*****  PERCENTS OF ROW TOTALS  -- TABLE  2

housing                satisfy
------                 ------
           low    medium    high    TOTAL
---------------------------------------------
tower      24.8     25.3    50.0 |   100.0
apt        35.4     25.1    39.5 |   100.0
atrium     26.8     33.1    40.2 |   100.0
terrace    48.0     26.7    25.3 |   100.0
--------------------------------|---------
TOTAL      33.7     26.5    39.7 |   100.0


MINIMUM ESTIMATED EXPECTED VALUE IS     63.41

STATISTIC                 VALUE    D.F.    PROB.
PEARSON CHISQUARE        60.286       6   0.0000
```

Similar tables are output for contact and influence.

Output W.2

Tower blocks: observed frequencies and goodness-of-fit statistics for log linear models. BMDP4F

```
***************************
* TABLE PARAGRAPH   3  *
***************************

*****  OBSERVED FREQUENCY TABLE  5

USING LEVEL   tower       OF VARIABLE   4   housing
              ********                      ********

influenc contact                satisfy
------   ------                 ------
                        low   medium   high   TOTAL
-----------------------------------------------------

low      low            21       21      28 |    70
         high           14       19      37 |    70
         ----------------------------------|--------
         TOTAL          35       40      65 |   140

medium   low            34       22      36 |    92
         high           17       23      40 |    80
         ----------------------------------|--------
         TOTAL          51       45      76 |   172

high     low            10       11      36 |    57
         high            3        5      23 |    31
         ----------------------------------|--------
         TOTAL          13       16      59 |    88

       TOTAL OF THE OBSERVED FREQUENCY TABLE IS      400

*****  ALL MODELS ARE REQUESTED--
```

MODEL	DF	LIKELIHOOD-RATIO CHISQ	PROB.	PEARSON CHISQ	PROB.	ITERATIONS
s.	15	60.55	0.0000	55.50	0.0000	1
c.	16	104.07	0.0000	95.46	0.0000	1
i.	15	79.55	0.0000	82.53	0.0000	1
s,c.	14	56.93	0.0000	52.27	0.0000	1
c,i.	14	75.94	0.0000	80.02	0.0000	1
i,s.	13	32.42	0.0021	31.65	0.0027	1
s,c,i.	12	28.80	0.0042	27.39	0.0068	1
sc.	12	50.19	0.0000	44.65	0.0000	1
si.	9	17.94	0.0359	17.53	0.0410	1
ci.	12	70.92	0.0000	74.68	0.0000	1
s,ci.	10	23.78	0.0082	23.68	0.0085	1
c,si.	8	14.32	0.0737	14.05	0.0806	1
i,sc.	10	22.06	0.0148	21.90	0.0156	1
sc,si.	6	7.58	0.2704	7.47	0.2797	1
si,ci.	6	9.30	0.1572	9.15	0.1653	1
ci,sc.	8	17.04	0.0297	16.74	0.0330	1
sc,si,ci.	4	0.57	0.9663	0.57	0.9663	4

Similar tables are output for apartments, atrium and terraced houses.

Output W.3

Tower blocks: fitted frequencies and standardized residuals from model containing S × I. BMDP4F

```
*****************
*  MODEL  1  *
*****************
```

MODEL	D.F.	LIKELIHOOD-RATIO CHI-SQUARE	PROB	PEARSON CHI-SQUARE	PROB
-----	----	----------	----	----------	----
si,c.	8	14.32	0.0737	14.05	0.0806

***** EXPECTED VALUES USING ABOVE MODEL

influenc	contact	satisfy			
------	------	------			
		low	medium	high	TOTAL
low	low	19.2	21.9	35.6	76.7
	high	15.8	18.1	29.4	63.3
	TOTAL	35.0	40.0	65.0	140.0
medium	low	27.9	24.6	41.6	94.2
	high	23.1	20.4	34.4	77.8
	TOTAL	51.0	45.0	76.0	172.0
high	low	7.1	8.8	32.3	48.2
	high	5.9	7.2	26.7	39.8
	TOTAL	13.0	16.0	59.0	88.0

***** STANDARDIZED DEVIATES = (OBS - EXP)/SQRT(EXP) FOR ABOVE MODEL

influenc	contact	satisfy		
------	------	------		
		low	medium	high
low	low	0.4	-0.2	-1.3
	high	-0.5	0.2	1.4
medium	low	1.2	-0.5	-0.9
	high	-1.3	0.6	1.0
high	low	1.1	0.8	0.7
	high	-1.2	-0.8	-0.7

Similar tables are output for apartments, atrium and terraced houses.

Example X   Educational plans of Wisconsin schoolboys

Description of data

Sewell and Shah (1968) have investigated for some Wisconsin highschool 'senior' boys and girls the relationship between variables:

(i) socioeconomic status (high, upper middle, lower middle, low);
(ii) intelligence (high, upper middle, lower middle, low);
(iii) parental encouragement (low, high);
(iv) plans for attending college (yes, no).

The data for boys are given in Table X.1.

Table X.1. Socioeconomic status, intelligence, parental encouragement and college plans for Wisconsin schoolboys

IQ	College plans	Parental encouragement	SES			
			L	LM	UM	H
L	Yes	Low	4	2	8	4
		High	13	27	47	39
	No	Low	349	232	166	48
		High	64	84	91	57
LM	Yes	Low	9	7	6	5
		High	33	64	74	123
	No	Low	207	201	120	47
		High	72	95	110	90
UM	Yes	Low	12	12	17	9
		High	38	93	148	224
	No	Low	126	115	92	41
		High	54	92	100	65
H	Yes	Low	10	17	6	8
		High	49	119	198	414
	No	Low	67	79	42	17
		High	43	59	73	54

The analysis

The binary variable college plans (CP) is treated as the observed response and we study its dependence upon socioeconomic status (SES), intelligence (IQ) and parental encouragement (PE) treated as explanatory variables on an equal footing.

For preliminary inspection we first calculate the percentages responding 'yes' to CP in each of the $4 \times 4 \times 2$ cells. This is done using BMDP4F.

We could then consider analysing these percentages directly using, say, BMDP2V or BMDP4V for analysis of variance, but given that the percentages vary from 1% to 88%, and that they are based on very unequal numbers, it is preferable to proceed instead by fitting a logistic model, for which we use BMDPLR.

Note that BMDP4F used above will fit loglinear models. Provided appropriate interaction terms are included so as to give the correct marginal fitted frequencies, loglinear models and logistic models are equivalent except that differences of parameter estimates are required for interpretation of the results if a loglinear model is fitted.

Instructions for both approaches are given, i.e. logistic models fitted by BMDPLR and equivalent loglinear models by BMDP4F.

Program

BMDP4F is used to compute the preliminary percentages responding 'yes' to CP. Various options for the format of the input data are available with BMDP4F; that chosen here is suitable also for BMDPLR. The following instructions print the observed frequencies set out as in Table X.1, and the required percentages.

```
/problem     title = 'example x.
             wisconsin schoolboys.  bmdp4f'.
/input       variables = 5.  format = free.
/variable    names = iq, cp, pe, ses, n.
/category    codes(ses,iq) = 1 to 4.
             names(ses,iq) = l, lm, um, h.    labels for output
             codes(pe) = 1, 2.
             names(pe) = low, high.
             codes(cp) = 1, 2.
             names(cp) = yes, no.
/table       indices = ses, pe, cp, iq.       print observed frequency table
             count = n.
/table       indices = cp, ses, pe, iq.
             count = n.
/print       percent = row.                   print percentages for CP
/end
1 1 1 1 4                                      64 lines of data
1 1 1 2 2                                      Code: cols 1, 4. L=1, LM=2, UM=3,
                                                           H=4
                                                     col. 2. Yes=1, No=2.
                                                     col. 3. Low=1, High=2.
. . .
```

The second **table** paragraph rearranges the data to make rows correspond to CP in order to print percentages for CP. Output X.1 shows the rearranged data and percentages.

BMDPLR is a stepwise logistic regression program. The following instructions carry out a stepwise procedure selecting terms from the main effects IQ, PE, SES; no interaction terms will be included. The stepwise aspect may be suppressed, to fit directly a model containing the three main effects, by inserting **model = iq, pe, ses.** into the **regress** paragraph; in principle this would be preferable, although with the given data each main effect is significant, so the stepwise procedure fits the full model.

```
/problem      title = 'example x.
              wisconsin schoolboys. bmdplr'.
/input        variables = 5. format = free.
/variable     names = iq, cp, pe, ses, n.
/group        codes(ses,iq) = 1 to 4.              labels for output
              names(ses,iq) = l, lm, um, h.
              codes(pe) = 1, 2.
              names(pe) = low, high.
              codes(cp) = 1, 2.
              names(cp) = yes, no.
/regress      dependent = cp.  count = n.
/print        cells = use.  sort = var.           print observed and predicted
/end                                               proportions in order of IQ, PE, SES.
1 1 1 1 4                                          64 lines of data as before
1 1 1 2 2
  . . .
```

The program generates design variables to represent the four levels of IQ and SES, and the two levels of PE as shown in Output X.2. Variables enter the model in the order PE, IQ, SES. Estimates of the coefficients when all three variables are included are given in Output X.3; also shown is the goodness-of-fit statistic, 2 log (maximized likelihood) = 25.236 with 24 d.f., agreeing with the result in *App. Stat.*

The lowest level L of SES is represented by the combination of estimated parameters equal to $0.252 - 0.054 - 0.806 = -0.608$; likewise we get -0.252, 0.054, 0.806 for LM, UM and H. These agree with those given in *App. Stat.* except there, instead of the mean being zero, the first level is set to zero.

Output X.4 shows the predicted probabilities of responding 'yes' to CP, listed in the order of IQ, PE, SES. They correspond to those in Table X.2, *App. Stat.*

To fit a logistic model containing interaction terms add a **model** statement to the **regress** paragraph specifying all the terms to be considered. Hierarchical models are implied. For example,

model = ses∗iq∗pe.

will include all possible interactions and main effects;

model = ses, iq∗pe.

will include all main effects plus the $IQ \times PE$ interaction.

BMDP4F can be used to achieve the same results by fitting equivalent loglinear models, instead of logistic models. For the equivalent of a main-effects logistic model containing SES, IQ and PE, we specify a loglinear model containing the three-factor interaction SES × IQ × PE (and lower-order terms in the hierarchical model) plus the interactions of the main effects with CP. This can be done by adding a **fit** paragraph to the BMDP4F program used above for computing observed percentages as follows. Identify variables by their initial letters: s = SES, i = IQ, p = PE, c = CP. Replace the **print** paragraph in the BMDP4F program with the following **fit** and **print** paragraphs:

```
/fit          model = cs, cp, ci, spi.        terms in loglinear model
/print        percent = row. expected.        print percentages for CP
              lambda. variance.               print expected frequencies
                                              parameter estimates
                                              variances/covariances of parameter esti-
                                                mates
/end
1 1 1 1 4
1 1 1 2 2
```

The output includes that given earlier (percentages for CP) plus goodness-of-fit statistics and fitted frequencies as in Output X.5, and estimates of all parameters in the model. Estimates for the SES × CP interaction are shown in Output X.6. Note that $-2 \log$ (likelihood ratio) = 25.24 with 24 d.f., agreeing with the logistic model value. Also, differing the estimated parameters for the SES × CP interaction across CP gives values for the four categories of SES equal to

$$-0.608, \; -0.252, \; 0.054, \; 0.806$$

agreeing exactly with those obtained using BMDPLR to fit the logistic model. Standard errors for these differences can be obtained from the output variance–covariance matrix of the parameter estimates (not shown here). Estimates for the categories of IQ and PE can likewise be obtained.

Suggested further work

Use BMDPLR to fit a main-effects logistic model using scores (e.g. 1, 2, 3 and 4) to represent the categories of SES and IQ. To do this insert **interval = ses, iq.** in the **regress** paragraph. Compare the fit of this model with that already fitted.

Output X.1

Observed frequencies and percentages responding 'yes' and 'no' to college plans. BMDP4F

***** OBSERVED FREQUENCY

iq	pe	ses	cp yes	no	TOTAL
I	low	l	4	349 |	353
		lm	2	232 |	234
		um	8	166 |	174
		h	4	48 |	52
		TOTAL	18	795 |	813
	high	l	13	64 |	77
		lm	27	84 |	111
		um	47	91 |	138
		h	39	57 |	96
		TOTAL	126	296 |	422

***** PERCENTS OF ROW TOTALS

iq	pe	ses	cp yes	no	TOTAL
I	low	l	1.1	98.9 |	100.0
		lm	0.9	99.1 |	100.0
		um	4.6	95.4 |	100.0
		h	7.7	92.3 |	100.0
		TOTAL	2.2	97.8 |	100.0
	high	l	16.9	83.1 |	100.0
		lm	24.3	75.7 |	100.0
		um	34.1	65.9 |	100.0
		h	40.6	59.4 |	100.0
		TOTAL	29.9	70.1 |	100.0

Similar tables are output for IQ=LM, UM, H.

Output X.2

Design variables for IQ, PE and SES. BMDPLR

VARIABLE NO. N A M E	VALUE OR INTERVAL	FREQ	DESIGN VARIABLES (1)	(2)	(3)
1 iq	1	1235	-1	-1	-1
	2	1263	1	0	0
	3	1238	0	1	0
	4	1255	0	0	1
3 pe	1	2085	-1		
	2	2906	1		
4 ses	1	1150	-1	-1	-1
	2	1298	1	0	0
	3	1298	0	1	0
	4	1245	0	0	1

Output X.3

Logistic model. Goodness-of-fit and estimates of parameters. BMDPLR

GOODNESS OF FIT CHI-SQ	(2*O*LN(O/E)) =	25.236 D.F.= 24 P-VALUE= 0.393
GOODNESS OF FIT CHI-SQ	(D. HOSMER) =	12.508 D.F.= 8 P-VALUE= 0.130
GOODNESS OF FIT CHI-SQ	(C.C.BROWN) =	10.906 D.F.= 2 P-VALUE= 0.004

TERM		COEFFICIENT	STANDARD ERROR	COEFF/S.E.
iq	(1)	-0.37930	0.6576E-01	-5.768
	(2)	0.35981	0.6197E-01	5.806
	(3)	0.99292	0.6336E-01	15.67
pe		1.2277	0.5070E-01	24.22
ses	(1)	-0.25213	0.6602E-01	-3.819
	(2)	0.54309E-01	0.6227E-01	0.8722
	(3)	0.80590	0.6344E-01	12.70
CONSTANT		-1.2162	0.5165E-01	-23.55

Output X.4

Logistic model. Observed proportions and fitted probabilities. BMDPLR

NUMBER yes	NUMBER no	OBSERVED PROPORTION yes	PREDICTED PROB.OF yes	S.E. OF PREDICTED PROB.	OBS-PRED -------- S.E.RES.	PRED. LOG ODDS	CHI	DEVIANCE	HAT MATRIX DIAGONAL	INFLUENCE	iq	pe	ses
4	349	0.0113	0.0175	0.0026	-0.9568	-4.025	-0.8888	-0.9499	0.1371	0.1454	1.00	1.00	1.00
2	232	0.0085	0.0249	0.0034	-1.7017	-3.670	-1.6024	-1.8509	0.1133	0.3699	1.00	1.00	2.00
8	166	0.0460	0.0335	0.0045	0.9729	-3.363	0.9173	0.8693	0.1111	0.1183	1.00	1.00	3.00
4	48	0.0769	0.0684	0.0092	0.2523	-2.611	0.2434	0.2389	0.0697	0.0048	1.00	1.00	4.00
13	64	0.1688	0.1722	0.0193	-0.0879	-1.570	-0.0785	-0.0787	0.2018	0.0020	1.00	2.00	1.00
27	84	0.2432	0.2290	0.0213	0.4231	-1.214	0.3577	0.3552	0.2852	0.0714	1.00	2.00	2.00
47	91	0.3406	0.2875	0.0234	1.7332	-0.908	1.3783	1.3555	0.3676	1.7466	1.00	2.00	3.00
39	57	0.4063	0.4611	0.0290	-1.3110	-0.156	-1.0774	-1.0816	0.3246	0.8260	1.00	2.00	4.00
9	207	0.0417	0.0313	0.0042	0.9330	-3.431	0.8720	0.8311	0.1265	0.1260	2.00	1.00	1.00
7	201	0.0337	0.0441	0.0054	-0.7957	-3.075	-0.7359	-0.7669	0.1447	0.1071	2.00	1.00	2.00
6	120	0.0476	0.0590	0.0072	-0.5782	-2.769	-0.5433	-0.5612	0.1172	0.0444	2.00	1.00	3.00
5	47	0.0962	0.1174	0.0136	-0.4996	-2.017	-0.4758	-0.4894	0.0927	0.0255	2.00	1.00	4.00
33	72	0.3143	0.2737	0.0237	1.1128	-0.976	0.9327	0.9195	0.2975	0.5244	2.00	2.00	1.00
64	95	0.4025	0.3498	0.0229	1.7534	-0.620	1.3941	1.3797	0.3678	1.7888	2.00	2.00	2.00
74	110	0.4022	0.4223	0.0233	-0.7173	-0.314	-0.5515	-0.5527	0.4089	0.3560	2.00	2.00	3.00
123	90	0.5775	0.6078	0.0225	-1.2245	0.438	-0.9067	-0.9029	0.4517	1.2351	2.00	2.00	4.00

12	126	0.0870	0.0634	0.0080	1.2292	-2.692	1.1337	1.0769	0.1494	0.2653	3.00	1.00	1.00
12	115	0.0945	0.0882	0.0099	0.2733	-2.336	0.2514	0.2487	0.1540	0.0136	3.00	1.00	2.00
17	92	0.1560	0.1161	0.0126	1.4247	-2.030	1.2989	1.2426	0.1688	0.4121	3.00	1.00	3.00
9	41	0.1800	0.2179	0.0217	-0.6984	-1.278	-0.6484	-0.6632	0.1380	0.0781	3.00	1.00	4.00
38	54	0.4130	0.4411	0.0288	-0.6512	-0.237	-0.5415	-0.5429	0.3086	0.1892	3.00	2.00	1.00
93	92	0.5027	0.5298	0.0239	-0.9711	0.119	-0.7371	-0.7365	0.4239	0.6938	3.00	2.00	2.00
148	100	0.5968	0.6048	0.0212	-0.3550	0.426	-0.2591	-0.2588	0.4672	0.1105	3.00	2.00	3.00
224	65	0.7751	0.7644	0.0164	0.5561	1.177	0.4265	0.4288	0.4324	0.2441	3.00	2.00	4.00
10	67	0.1299	0.1131	0.0137	0.5010	-2.059	0.4636	0.4542	0.1436	0.0421	4.00	1.00	1.00
17	79	0.1771	0.1541	0.0163	0.6969	-1.703	0.6251	0.6131	0.1956	0.1181	4.00	1.00	2.00
6	42	0.1250	0.1983	0.0200	-1.3592	-1.397	-1.2742	-1.3465	0.1211	0.2545	4.00	1.00	3.00
8	17	0.3200	0.3441	0.0289	-0.2662	-0.645	-0.2535	-0.2550	0.0927	0.0072	4.00	1.00	4.00
49	43	0.5326	0.5978	0.0282	-1.5298	0.396	-1.2752	-1.2659	0.3051	1.0277	4.00	2.00	1.00
119	59	0.6685	0.6797	0.0213	-0.4012	0.752	-0.3182	-0.3173	0.3707	0.0948	4.00	2.00	2.00
198	73	0.7306	0.7424	0.0176	-0.5939	1.059	-0.4447	-0.4425	0.4394	0.2764	4.00	2.00	3.00
414	54	0.8846	0.8594	0.0110	2.1606	1.810	1.5695	1.6122	0.4723	4.1787	4.00	2.00	4.00

Example X

Output X.5

Loglinear model. Goodness-of-fit and expected values. BMDP4F

MODEL	D.F.	LIKELIHOOD-RATIO CHI-SQUARE	PROB	PEARSON CHI-SQUARE	PROB	I
-----	----	----------	----	----------	----	-
cs,cp,ci,spi.	24	25.24	0.3930	24.44	0.4367	

***** EXPECTED VALUES USING ABOVE MODEL

iq	pe	ses	cp		
------	------	------	------		
			yes	no	TOTAL
I	low	I	6.2	346.8	353.0
		lm	5.8	228.2	234.0
		um	5.8	168.2	174.0
		h	3.6	48.4	52.0
		TOTAL	21.4	791.6	813.0
	high	I	13.3	63.7	77.0
		lm	25.4	85.6	111.0
		um	39.7	98.3	138.0
		h	44.3	51.7	96.0
		TOTAL	122.6	299.4	422.0

Similar tables are output for IQ = LM, UM, H.

Output X.6

Loglinear model. Estimates of parameters for SES × CP. BMDP4F

***** ESTIMATES OF THE LOG-LINEAR PARAMETERS (LAMBDA) IN THE MODEL ABOVE

ses	cp	
------	------	
	yes	no
I	-0.304	0.304
lm	-0.126	0.126
um	0.027	-0.027
h	0.403	-0.403

***** RATIO OF THE LOG-LINEAR PARAMETER ESTIMATE TO ITS STANDARD ERROR

ses	cp	
------	------	
	yes	no
I	-7.618	7.618
lm	-3.820	3.820
um	0.873	-0.873
h	12.703	-12.703

Similar tables are output for all other terms.

FURTHER SETS OF DATA

Set 3   Survival times of rats

Table S.3 (J.S. Maritz, personal communication) gives data from an experiment on carcinogenesis in rats. Eighty rats were divided at random into 4 groups of 20 rats each, and treated as follows:

Group I D, no I, no P;

II D, I, no P;

III D, no I, P;

IV D, I, P;

where D is thought to produce cancer, I is thought to act as an inhibitor and P is thought to accelerate the appearance of cancer. The data in Table S.3 are survival times in days; after 192 days the experiment was ended, and a post mortem was conducted on every surviving rat to assess the presence or absence of cancer. In the table, 192^- means that the rat survived 192 days but was found to have cancer. The superscript $^+$ means death from a cause unrelated to cancer; in particular, 192^+ means that on post mortem the rat did not have cancer.

Table S.3. Survival times in days for four groups of rats

Group I; D		Group II; DI		Group III; DP		Group IV; DIP	
18^+	106	2^+	192^+	37	51	18^+	127
57	108	2^+	192^+	38	51	19^+	134
63^+	133	2^+	192^+	42	55	40^+	148
67^+	159	2^+	192^+	43^+	57	56	186
69	166	5^+	192^+	43	59	64	192^+
73	171	55^+	192^+	43	62	78	192^+
80	188	78	192^+	43	66	106	192^+
87	192^-	78	192^-	43	69	106	192^+
87^+	192^-	96	192^-	48	86	106	192^+
94	192^-	152	192^-	49	177	127	192^+

The analysis

Various approaches can be taken, comparing the number of animals observed

153

to have cancer or their survival times. Here we concentrate solely on the time to death from cancer. Thus death from other causes is treated as censoring and the post mortem classification ignored.

Program BMDP1L uses the actuarial life table method (or, if requested, the slightly different product limit method) to estimate the survival distribution. This is done separately for each treatment group and the four distributions are plotted for comparison on the same graph. Median survival times and approximate standard errors are computed. There are marked differences in survival time.

A more precise comparison taking into account the effects of treatments I and P is obtained by using BMDP2L to fit a proportional hazards model (Cox and Oakes, 1984, Chapter 7). The hazard function for an individual at time t is

$$h(t, \mathbf{z}) = h_0(t) \exp(\boldsymbol{\beta}^T \mathbf{z})$$

where $\mathbf{z}$ is a vector of covariates representing treatments and $\boldsymbol{\beta}$ is a vector of unknown parameters; $h_0(t)$ is the hazard when $\mathbf{z} = \mathbf{0}$. Main effects and interactions of I and P are accounted for by writing $\mathbf{z} = (I, P, I \times P)^T$ with $I = -1, 1$ and $P = -1, 1$ to denote the two levels, absent and present, of treatments I and P, respectively. Output from BMDP2L includes estimates of the coefficients β, with approximate standard errors, and plots of the estimated survival distributions.

Program

The same data format is used for both BMDP1L and BMDP2L. Although the levels of I and P are not specifically required for BMDP1L, they can be used to generate a single code to identify the four treatment groups. The instructions for estimating the survival distribution by the actuarial life table method using BMDP1L are:

```
/problem      title = 'set3.  cancer experiment.
              bmdp1l.'.
/input        variables = 4.  format = free.
/variable     names = days, cancer, i, p.
              add = new.
/transform    k = i + 2*p.
/form         time = days.  status = cancer.
              response = 1.
/estimate     grouping = k.  period = 20.

/end
18  0 -1 -1
57  1 -1 -1

      . . .
```

k = treatment group code
name of survival measure
1 = death from cancer
no. of intervals for survival distribution

80 lines of data

Code: col. 2 no cancer/censored = 0,
        death from cancer = 1.
        col. 3 no $I = -1$, $I = 1$.
        col. 4 no $P = -1$, $P = 1$.

Note: Since the post mortem classification is ignored, both 192^- and 192^+ are coded as 0 in col. 2 to indicate censoring.

Output 3.1 shows the life table and survival analysis for group I. The median survival times (also shown in Output 3.1, for group I) are markedly different in the four groups, viz:

Group	I	II	III	IV
Median (days)	127.4	*	51.3	132.8
s.e.	24.1	*	4.0	9.9

The median in group II cannot be estimated due to the large number of censored observations but it is greater than 148.8 days, the 0.75 quartile (which is also output). Output 3.2 plots the survival distributions for groups I to IV.

BMDP2L fits a proportional hazards model. The instructions are:

```
/problem    title = 'set3. cancer experiment.
            bmdp2l'.
/input      variables = 4. format = free.
/variable   names = days, cancer, i, p.
            add = new.
/transform  ixp = i*p.                          generate I × P
/form       time = days. status = cancer.
            response = 1.
/regress    covariate = i, p, ixp.              main effects and interaction
/plot       type is surv.                       plot survival distribution
            pattern = -1, -1, 1.
            pattern = 1, -1, -1.
            pattern = -1, 1, -1.
            pattern = 1, 1, 1.
/print      survival.
/end
18 0 -1 -1
57 1 -1 -1                                       80 lines of data as before
```

...

Output 3.3 shows the estimated treatment coefficients. The main effects I and P are highly significant; the interaction $I \times P$ is less so. Under this model the estimated survival distribution is given by

$$\hat{S}(t, \mathbf{z}) = \hat{S}_0(t) \exp \boldsymbol{\beta}^T \mathbf{z},$$

where $\hat{S}(t)$ is the estimated survival distribution when $\mathbf{z} = 0$. Thus for group I, with $I = -1$, $P = -1$, $I \times P = 1$,

$$\hat{S}(t, \mathbf{z}) = \hat{S}_0(t) \exp(.8803 - .6988 - .1795)$$
$$= \hat{S}_0(t) \times 1.002$$

and the value 1.002, called the conversion factor, is shown (with that for groups II, III and IV) in Output 3.4. Values of $\hat{S}_0(t)$ and residuals are also output (not shown here). Output 3.4 plots $\hat{S}(t, \mathbf{z})$ for each of the groups; the distributions are similar to those in Output 3.2 but with a greater difference between groups I and IV.

Suggested further work

(i) Use BMDP2L to compare just groups I and IV under a proportional hazards model. (Add **use = ixp eq 1.** to the **transform** paragraph. Remember also to amend the **covariate** and **pattern** instructions.)

(ii) Consider ways of checking the adequacy of the proportional hazards model.

(iii) Compare the regression estimates in Output 3.3 with the estimates obtained directly by comparing medians.

(iv) What information is provided by the post mortem data and how could it be analysed?

Output 3.1

Life table and survival analysis. Treatment group I. BMDP1L

PROPORTION SURVIVING AT INTERVAL	ENTERED	WITHDRAWN	LOST	DEAD	EXPOSED	PROPORTION DEAD	PROPORTION SURVIVING	PROPORTION SURVIVING BEGINNING OF INTERVAL	HAZARD (S.E.)	DENSITY (S.E.)
0.00 - 9.60	20	0	0	0	20.0	0.0000	1.0000	1.0000 / 0.0000	0.0000 / 0.0000	0.0000 / 0.0000
9.60 - 19.20	20	1	0	0	19.5	0.0000	1.0000	1.0000 / 0.0000	0.0000 / 0.0000	0.0000 / 0.0000
19.20 - 28.80	19	0	0	0	19.0	0.0000	1.0000	1.0000 / 0.0000	0.0000 / 0.0000	0.0000 / 0.0000
28.80 - 38.40	19	0	0	0	19.0	0.0000	1.0000	1.0000 / 0.0000	0.0000 / 0.0000	0.0000 / 0.0000
38.40 - 48.00	19	0	0	0	19.0	0.0000	1.0000	1.0000 / 0.0000	0.0000 / 0.0000	0.0000 / 0.0000
48.00 - 57.60	19	0	0	1	19.0	0.0526	0.9474	1.0000 / 0.0000	0.0056 / 0.0056	0.0055 / 0.0053
57.60 - 67.20	18	2	0	0	17.0	0.0000	1.0000	0.9474 / 0.0512	0.0000 / 0.0000	0.0000 / 0.0000
67.20 - 76.80	16	0	0	2	16.0	0.1250	0.8750	0.9474 / 0.0512	0.0139 / 0.0098	0.0123 / 0.0082
76.80 - 86.40	14	0	0	1	14.0	0.0714	0.9286	0.8289 / 0.0902	0.0077 / 0.0077	0.0062 / 0.0060
86.40 - 96.00	13	1	0	2	12.5	0.1600	0.8400	0.7697 / 0.1014	0.0181 / 0.0128	0.0128 / 0.0085
96.00 - 105.60	10	0	0	0	10.0	0.0000	1.0000	0.6466 / 0.1167	0.0000 / 0.0000	0.0000 / 0.0000

Output 3.1 (continued)

PROPORTION SURVIVING AT INTERVAL	ENTERED	WITHDRAWN	LOST	DEAD	EXPOSED	PROPORTION DEAD	PROPORTION SURVIVING	PROPORTION SURVIVING AT BEGINNING OF INTERVAL	HAZARD (S.E.)	DENSITY (S.E.)
105.60 - 115.20	10	0	0	2	10.0	0.2000	0.8000	0.6466 / 0.1167	0.0231 / 0.0163	0.0135 / 0.0089
115.20 - 124.80	8	0	0	0	8.0	0.0000	1.0000	0.5173 / 0.1241	0.0000 / 0.0000	0.0000 / 0.0000
124.80 - 134.40	8	0	0	1	8.0	0.1250	0.8750	0.5173 / 0.1241	0.0139 / 0.0139	0.0067 / 0.0065
134.40 - 144.00	7	0	0	0	7.0	0.0000	1.0000	0.4526 / 0.1243	0.0000 / 0.0000	0.0000 / 0.0000
144.00 - 153.60	7	0	0	0	7.0	0.0000	1.0000	0.4526 / 0.1243	0.0000 / 0.0000	0.0000 / 0.0000
153.60 - 163.20	7	0	0	1	7.0	0.1429	0.8571	0.4526 / 0.1243	0.0160 / 0.0160	0.0067 / 0.0065
163.20 - 172.80	6	0	0	2	6.0	0.3333	0.6667	0.3879 / 0.1222	0.0417 / 0.0289	0.0135 / 0.0089
172.80 - 182.40	4	0	0	0	4.0	0.0000	1.0000	0.2586 / 0.1105	0.0000 / 0.0000	0.0000 / 0.0000
182.40 - 192.00	4	3	0	1	2.5	0.4000	0.6000	0.2586 / 0.1105	0.0521 / 0.0504	0.0108 / 0.0095
								0.1552 / 0.1040 (AT END OF LAST INTERVAL)		

QUANTILE	ESTIMATE	STANDARD ERROR
75TH	87.94	9.89
MEDIAN (50TH)	127.36	24.08
25TH	183.20	17.55

Similar tables are output for treatment groups II, III and IV.

Output 3.2

Survival distributions for groups I, II, III and IV. BMDP1L

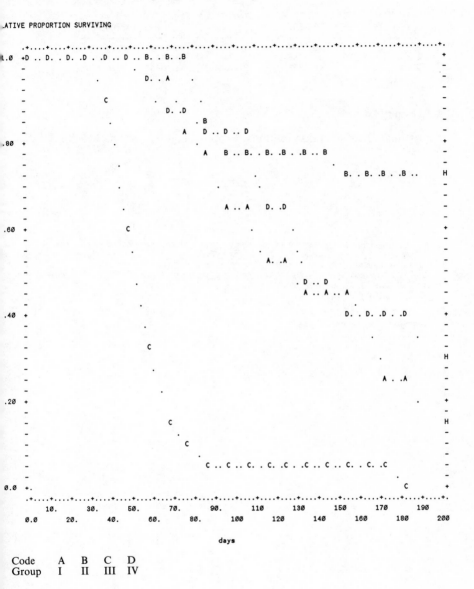

```
_ATIVE PROPORTION SURVIVING

    .+....+....+....+....+....+....+....+....+....+....+....+....+....+....+....+....+....+....+....+.
1.0  +D .. D. . D. .D . .D .. D .. B. . B. .B                                                      +
     -                         .             D. . A   .                                           -
     -                    C          .   .   .                                                    -
     -                            D. .D                                                           -
     -                    .           . B                                                         -
     -                           A   D .. D .. D                                                  -
 .80 +                   .                                                                        +
     -                            A    B .. B. . B. .B . .B .. B                                  -
     -              .                                    B. . B. .B . .B ..          H            -
     -                   .                       .    .                                           -
     -                   .        A .. A    D. .D                                                 -
 .60 +              C                           .      .                                          +
     -                   .                                                                        -
     -                                    A. .A   .                                               -
     -              .                       . D .. D                                              -
     -                                      A .. A .. A                                           -
 .40 +                   .                       D. . D. .D . .D                                  +
     -                   .                                   .         .            -
     -              C                                                                             -
     -              .                                          .                    H            -
     -                   .                            A . .A                        -
 .20 +                                                                    .         +
     -                   C                                                          H            -
     -              .    C                                                          -
     -         C .. C .. C. . C. .C . .C .. C .. C. . C. .C                         -
 0.0 +.                                                           C                 +
     .+....+....+....+....+....+....+....+....+....+....+....+....+....+....+....+....+....+....+....+.
        10.       30.       50.       70.       90.      110      130      150      170      190
     0.0       20.       40.       60.       80.       100      120      140      160      180      200

                                        days
```

Code	A	B	C	D
Group	I	II	III	IV

Output 3.3

Estimated coefficients of I, P and $I \times P$. BMDP2L

VARIABLE	COEFFICIENT	STANDARD ERROR	COEFF./S.E.	EXP(COEFF.)
3 i	-0.8803	0.1770	-4.9744	0.4146
4 p	0.6988	0.1742	4.0114	2.0114
5 IXP	-0.1795	0.1733	-1.0360	0.8357

Output 3.4

Estimated survival distributions. Proportional hazards model. BMDP2L

PATTERN	CONVERSION FACTOR **	3 i	4 p	5 IXP
1	1.002	-1	-1	1
2	0.247	1	-1	-1
3	5.805	-1	1	-1
4	0.697	1	1	1

** USE THE CONVERSION FACTOR AS AN EXPONENT TO CONVERT THE
SURVIVAL ESTIMATE FOR THE MEAN COVARIATE VECTOR TO THE
SURVIVAL PROBABILITY FOR A PARTICULAR COVARIATE PATTERN.
THE PROPORTIONAL HAZARDS SURVIVAL FOR THE MEAN COVARIATE
VECTOR IS PRINTED WHEN REQUESTED WITH THE SURVIVAL OPTION
IN THE PRINT PARAGRAPH.

PATTERN	SYMBOL	PAGE
1	A	8
2	B	8
3	C	8
4	D	8

Output 3.4 (continued)

ED SURVIVAL FUNCTION

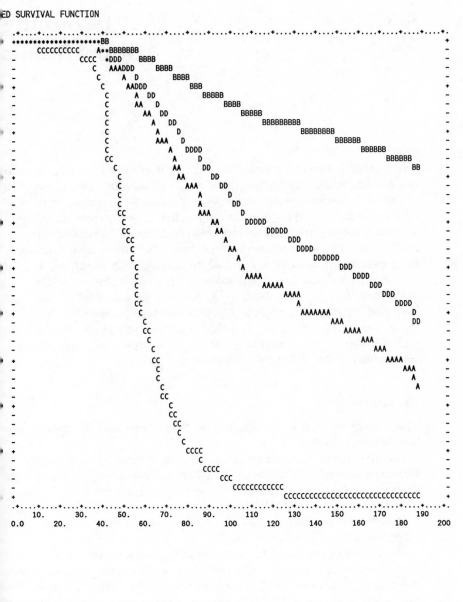

Set 14

Grouping of perfumes

Ten nominally similar perfumes for use in a disinfectant were assessed as follows. Ten small measured samples of the disinfectant, differing only in respect of the perfume, were placed in ten identical containers. A panel of 30 judges were each asked to smell the samples and to group them according to whether or not any difference in smell could be detected; the number of groups could be any number from one (no detectable difference) to ten (all different). Each panel member was also asked to classify each group of samples according to acceptability of perfume into 'like', 'indifferent' or 'dislike'.

The data are given in Table S.14. Samples judged to smell alike are bracketed together; thus, the first judge allocated the 10 samples into 4 groups, the second into 8 groups, etc. The identification of any groups in which the perfumes appear to be indistinguishable is of importance, and also the acceptability of the perfume in these groups.

The analysis

The primary object is to see whether the ten samples may be grouped into homogeneous clusters.

The first step is to decide on some measure of association, or similarity, which may be used as a basis for cluster analysis. We take simply the number of times that each pair of samples was grouped together; see Fig. S.14. This array is then analysed using BMDP1M, a single linkage cluster program.

Once clusters have been identified it requires only a simple analysis to check which clusters are 'liked' or 'disliked'.

An alternative procedure to the single linkage cluster analysis is metrical multidimensional scaling. A configuration in a small number of dimensions is determined by finding the largest eigenvalues of the similarity matrix, this having first been transformed into a distance matrix and 'centred' (Mardia *et al.*, 1979, Chapter 14). The eigenvalues corresponding to the largest eigenvalues are taken in pairs, giving coordinates which can be plotted to provide a visual display of the configuration. Provided the number of dimensions is small, ideally not more than two, this is an effective way to identify clusters.

162

Table S.14. Allocation by thirty judges of ten perfumes into groups

Judge	Acceptability		
	Like	Indifferent	Dislike
1	(2, 5, 8) (1, 4, 6, 9, 10)		(3) (7)
2	(1) (6) (7)	(3)	(2, 4, 5) (8) (9) (10)
3	(1, 2, 4, 5, 6, 9, 10)	(8)	(3) (7)
4	(3, 7, 10)		(5, 6, 8) (1, 2, 4, 9)
5	(5) (6) (9)	(1) (2) (4) (7) (8)	(3) (10)
6	(1, 4, 7, 8) (5) (6) (9) (10)		(2) (3)
7	(7)	(5) (4, 8, 9, 10)	(3) (1, 2, 6)
8		(1) (4) (5) (6) (10)	(2, 3, 9) (7, 8)
9	(1, 4, 5, 7, 8, 10) (2, 6, 9)		(3)
10	(1, 2, 5, 6) (7, 8, 9, 10)		(3, 4)
11	(1, 4, 8, 9, 10) (2, 5, 6, 7)		(3)
12	(8) (4, 6) (9, 10)	(5) (7) (1, 2)	(3)
13	(2, 7, 9) (5, 8, 10)	(1) (4, 6)	(3)
14		(2) (5) (7) (9)	(1, 4, 8, 10) (3, 6)
15	(1, 4, 9) (2, 5, 6, 7, 8, 10)		(3)
16	(1, 4, 10)	(2) (5) (6) (7) (8) (9)	(3)
17	(5, 6, 8)		(1, 2, 7, 9, 10) (3, 4)
18	(1, 4, 7, 9)	(2, 5, 6, 8, 10)	(3)
19	(2, 3, 8)	(5, 6, 10)	(1, 4, 7, 9)
20	(1) (2) (4) (5) (6) (7) (8) (9) (10)		(3)
21	(2, 4, 6, 9) (8, 10)	(5, 7) (1)	(3)
22	(2, 6, 8, 10)	(1, 4, 5, 9)	(3) (7)
23	(4, 9) (6, 10)		(3, 7) (1, 2, 5, 8)
24	(1, 9, 10)	(2, 5) (4, 6, 8)	(3, 7)
25		(7) (1, 2, 5) (4, 6, 8, 9, 10)	(3)
26	(5, 10)	(1, 2, 6, 9)	(3, 7) (4, 8)
27	(1, 4, 7, 9)	(2, 5, 6, 8, 10)	(3)
28	(1, 2, 6)	(5, 9, 10)	(3, 4, 7, 8)
29	(2) (3, 5, 6, 10)		(1, 4, 7, 8, 9)
30	(1, 4)	(3, 5)	(2, 7) (6, 8) (9, 10)

We concentrate here on using the single linkage cluster program BMDP1M. The multidimensional scaling approach is suggested as further work (see later).

Sample

1 2 3 4 5 6 7 8 9 10

	1	2	3	4	5	6	7	8	9	10
1	30	12	0	18	8	9	9	8	14	10
2	12	30	3	6	14	14	8	10	11	7
3	0	3	30	3	2	2	5	2	1	3
4	18	6	3	30	6	9	8	12	15	10
Sample 5	8	14	2	6	30	15	6	11	8	13
6	9	14	2	9	15	30	4	11	10	13
7	9	8	5	8	6	4	30	9	9	5
8	8	10	2	12	11	11	9	30	8	14
9	14	11	1	15	8	10	9	8	30	14
10	10	7	3	10	13	13	5	14	14	30

Fig. S.14. Similarity matrix: the number of times
each pair of samples is grouped together

Program

The instructions for BMDP1M are as follows:

```
/problem      title = 'set 14 perfumes. bmdp1m.'.
/input        variables = 10.  type = simi.          similarity matrix input
              format = free.
/variable     names = sample1, sample2, sample3,
              sample4, sample5, sample6, sample7,
              sample8, sample9, sample10.
/end
30 12 0 18 8 9 9 8 14 10                              10 lines of data as in Fig. S.14
12 30 3 6 14  14 8 10 11 7
```

Initially each sample is considered as a separate cluster. Thereafter, starting
with one sample, further ones are linked, one at a time, such that at each stage
the two clusters so joined have maximum similarity. Output 3.1 gives a
summary table of the clusters formed and Output 3.2 shows a step-by-step tree
diagram of how the clusters are linked together.

Samples 1 and 4 are linked, then samples 1, 4 and 9, etc. The horizontal and
diagonal lines in Output 3.2 mark the clusters. Suggested clusters are: samples
1, 4, 8, 9, 10; samples 2, 5, 6; sample 7; sample 3.

Suggested further work

(i) Examine which of the identified clusters are 'liked' or 'disliked'.
(ii) Instead of BMDP1M, try using the multidimensional scaling approach to
determine clusters. First transform the similarity matrix $S = (s_{ij})$ to a
distance matrix $D = (d_{ij})$ where

$$d_{ij}^2 = s_{ii} - 2s_{ij} + s_{jj}.$$

Then construct $\mathbf{A} = (a_{ij}) = (-\frac{1}{2}d_{ij}^2)$ and $\mathbf{B} = (b_{ij})$ where

$$b_{ij} = a_{ij} - \bar{a}_{i.} - \bar{a}_{.j} + \bar{a}_{..};$$

$\bar{a}_{i.}, \bar{a}_{.j}$ and $\bar{a}_{..}$ are row, column and overall means of (a_{ij}). BMDP4M can be used for a principal component analysis of $\mathbf{B}$. This procedure is sometimes called the classical scaling method.

Output 14.1

Summary table of clusters. BMDP1M

VARIABLE NAME	NO.	OTHER BOUNDARY OF CLUSTER	NUMBER OF ITEMS IN CLUSTER	DISTANCE OR SIMILARITY WHEN CLUSTER FORMED
sample1	1	3	10	5.00
sample4	4	1	2	18.00
sample9	9	1	3	15.00
sample8	8	10	2	14.00
sample10	10	1	5	14.00
sample2	2	6	3	14.00
sample5	5	6	2	15.00
sample6	6	1	8	13.00
sample7	7	1	9	9.00
sample3	3	1	10	5.00

Output 14.2

Step-by-step tree diagram of clusters. BMDP1M

```
TREE PRINTED OVER SIMILARITY MATRIX (SCALED 0-100).
CLUSTERING BY MINIMUM DISTANCE METHOD.
   VARIABLE
NAME    NO.
        ---------------------------/
sample1 (  1) 59/46/26 33/39 26 29/29/ 0/
             / /    /      / / / /
            / /    /      / / / /
sample4 (  4)/49/39 33/19 19 29/26/ 9/
               /    /      / / / /
              /    /      / / / /
sample9 (  9)/26 46/36 26 33/29/ 3/
             /    /      / / / /
          ----/      / / / /
sample8 (  8) 46/33 36 36/29/ 6/
             /      / / / /
            /      / / / /
sample10( 10)/23 43 43/16/ 9/
            /      / / /
        -------/    / / /
sample2 (  2) 46 46/26/ 9/
               /    / /
          ----/    / /
sample5 (  5) 49/19/ 6/
             /    / /
            /    / /
sample6 (  6)/13/ 6/
             /  /
            /  /
sample7 (  7)/16/
             /
            /
sample3 (  3)/
```

```
THE VALUES IN THIS TREE HAVE BEEN SCALED 0 TO 100
ACCORDING TO THE FOLLOWING TABLE
```

VALUE ABOVE	SIMILARITY	VALUE ABOVE	SIMILARITY
0	0.000	50	15.000
5	1.500	55	16.500
10	3.000	60	18.000
15	4.500	65	19.500
20	6.000	70	21.000
25	7.500	75	22.500
30	9.000	80	24.0C0
35	10.500	85	25.500
40	12.000	90	27.000
45	13.500	95	28.500

References

The abbreviation *App. Stat.* used frequently throughout this handbook refers to Cox and Snell (1981), referenced below.

Baxter, G.P. and Landstredt, O.W. (1940). A revision of the atomic weight of iodine. *J. Amer. Chem. Soc.*, **62**, 1829–34.

Biggers, J.D. and Heyner, S. (1961). Studies on the amino acid requirements of cartilaginous long bone rudiments *in vitro*. *J. Exp. Zool.*, **147**, 95–112.

Box, G.E.P. and Cox, D.R. (1964). An analysis of transformations (with discussion). *J. R. Statist. Soc.*, B, **26**, 211–52.

Brownlee, K.A. (1965). *Statistical Theory and Methodology in Science and Engineering.* 2nd edn. New York: Wiley.

Cox, D.R. (1958). *Planning of Experiments.* New York: Wiley.

Cox, D.R. and Oakes, D. (1984). *Analysis of Survival Data.* London: Chapman and Hall.

Cox, D.R. and Snell, E.J. (1981). *Applied Statistics. Principles and Examples.* London: Chapman and Hall.

Desmond, D.J. (1954). Quality control on the setting of voltage regulators. *Applied Statist.*, **3**, 65–73.

Dixon, W.J., Brown, M.B., Engelman, L., Frane, J.W., Hill, M., Jennrich, R.I. and Toposek, J.D. (1985). *BMDP Statistical Software.* University of California Press. (This is a reprint of the 1983 edn.)

Fedorov, V.D., Maximov, V.N. and Bogorov, V.G. (1968). Experimental development of nutritive media for micro-organisms. *Biometrika*, **55**, 43–51.

Feigl, P. and Zelen, M. (1965). Estimation of exponential survival probabilities with concomitant information. *Biometrics*, **21**, 826–38.

Greenberg, R.A. and White, C. (1963). The sequence of sexes in human families. Paper presented to the 5th International Biometric Conference, Cambridge.

Hill, M. (1984). *BMDP User's Digest. A condensed guide to the BMDP computer programs.* University of California Press.

John, J.A. and Quenouille, M.H. (1977). *Experiments: Design and Analysis*, 2nd edn. London: Griffin.

Johnson, N.L. (1967). Analysis of a factorial experiment. (Partially confounded 2^3). *Technometrics*, **9**, 167–70.

Ling, R.F. (1984). Review of *Applied Statistics. Principles and applications*. Cox and Snell (1981). *J. Amer. Statist. Ass.*, **79**, 229–31.

MacGregor, G.A., Markandu, N.D., Roulston, J.E. and Jones, J.C. (1979). Essential hypertension: effect of an oral inhibitor of angiotension-converting enzyme. *Brit. Med. J.*, **2**, 1106–9.

Madsen, M. (1976). Statistical analysis of multiple contingency tables. Two examples. *Scand. J. Statist.*, **3**, 97–106.

Mardia, K.V., Kent, J.T. and Bibby, J.M. (1979). *Multivariate Analysis*, London: Academic Press.

Mooz, W.E. (1978). Cost analysis of light water reactor power plants. *Report R-2304-DOE*. Rand Corp., Santa Monica, Calif.

Morton, A.Q. (1965). The authorship of Greek prose (with discussion). *J. R. Statist. Soc.*, A, **128**, 169–233.

Proschan, F. (1963). Theoretical explanation of observed decreasing failure rate. *Technometrics*, **5**, 375–83.

Ries, P.N. and Smith, H. (1963). The use of chi-square for preference testing in multidimensional problems. *Chem. Eng. Progress*, **59**, 39–43.

Sewell, W.H. and Shah, V.P. (1968). Social class, parental encouragement and educational aspirations. *Amer. J. Sociol.*, **73**, 559–72.

Woolf, B. (1955). On estimating the relation between blood group and disease. *Ann. Hum. Genetics*, **19**, 251–3.

Index

add, 8
analysis, 68, 73
Analysis of variance, 5, 6, 64, 68, 69, 75, 80, 85, 88, 95, 100, 105, 112, 115, 117, 143

Backward selection, see Regression
Balanced incomplete block
  between block analysis, 57
  design, 56
  within block analysis, 56
between, 74
bform, 74
Binary variable, 143
  see also Logistic model
BMDP
  language, 7
  1D, 5, 19, 23
  2D, 5, 14
  6D, 5, 19, 32
  7D, 5, 104, 105
  9D, 5, 63, 68, 86, 87, 88
  4F, 5, 86, 88, 137, 143, 145
  1L, 5, 154
  2L, 5, 154, 155
  1M, 5, 163, 164
  4M, 5, 6, 165
  1R, 5, 6, 7, 27, 35, 40, 57, 63, 69, 75, 79, 98
  2R, 5, 6, 7, 27, 35, 40, 41, 58, 64, 69, 75, 80
  3R, 5, 6, 7, 51, 127, 128
  9R, 5, 6, 42, 100, 101
  AR, 5, 6, 7, 53, 122, 128
  LR, 5, 6, 51, 52, 74, 80, 90, 133, 143, 144
  2V, 5, 6, 9, 69, 75, 88, 89, 95, 106, 113, 115

3V, 5, 6, 106
4V, 5, 6, 68, 73, 90
8V, 5, 6, 9, 69, 90, 104, 105, 112

Case control studies, 133
cell, 138
Cluster analysis, 3, 5, 162
Component of variance, 104, 115
Computation, see Error; Tolerance
Confounding, 79
Consistency, 23, 121
Contrast, 79, 80
  see also Orthogonal polynomial
covergence, 51
Correlation matrix, 57, 58, 100
count, 16

Data
  set, 8
  file, 9
  format, 9
days, 14, 15
design, 73
Design variable 133, 144
Distance matrix, 162, 164

Empty cell, 94, 95
end, 8
Error, 116
Example
  A, 5, 13–17
  B, 5, 18–21
  C, 3, 5, 22–25
  D, 3, 5, 26–30
  E, 5, 31–33
  F, 5, 8, 34–37
  G, 5, 38–49, 61
  H, 3, 5, 50–55, 128

169